# Organic Farming for Sustainable Horticulture

# Organic Farming for Sustainable Horticulture

R.K. Pawar

Oxford Book Company
Jaipur ▪ New Delhi

*Published By*

**Oxford** Book Company

**Regd. Off.** "Bony Residency", Gate No. 2
Opp. Tilak Public School, Vishweswariya Nagar,
Gopalpura Road, Jaipur - 302018, India
Ph: 0141 - 2761280, Telefax: 2761381
email - oxfordbookjpr@gmail.com
website - oxfordbookcompany.com

**Delhi Office:** 102, 1st Floor, Satyam House,
4327/3, Ansari Road, Darya Ganj,
New Delhi - 110002, India
Ph: 011-45652440

Published in India, 2025



ISBN : 9788189473730

Price : ₹ 1885/-



Typeset by Shivangi Computers, Jaipur

Printed and bound in India by Thomson Press Ltd.,
New Delhi

# Preface

The incentive to practice organic farming has been found in the emerging desire for wholesome agriculture products and the anxiety generated by scientific studies which link pesticides plant growth regulators and other chemical agents with cancer and other agents. In such a scenario horticulture, the science and art of growing fruits, vegetables ornamental trees, bushes and flowers, has not been ignored. Excluding the use of synthetic fertilizers and chemicals, and incorporating the use of sustainable, eco-friendly farm methods, organic farming is the latest trend emerging in horticulture, projecting immense scope and potential.

The present text has been designed as an introductory manual which serves to acquaint readers with the prospects offered by organic farming in horticulture. Elaborating upon its content based upon extensive market research and analysis of organic farming the book makes an assessment of the factors which are critical to the success of organic horticulture, especially as never techniques of horticulture farming, based on the latest scientific procedures, are being introduced. In addition, the book incorporates practices which are surrounding the field, taking care to ensure that critical perspectives are included which offer insight into the advantages, challenges sustainable horticultural practice.

*R.K. Pawar*

# Contents


| | | |
|---|---|---|
| | *Preface* | v |
| 1. | Sustainable Horticulture | 1 |
| 2. | Organic Farming and Biodiversity | 25 |
| 3. | Organic Farming and Gardening Systems and Approaches | 34 |
| 4. | Techniques of Organic Gardening and Farming | 61 |
| 5. | Succession Planting | 86 |
| 6. | Motivations for Organic Agriculture | 108 |
| 7. | Acquisition and Adoption of Organic Farming Practices | 134 |
| | *Bibliography* | 157 |
| | *Index* | 159 |

# 1

# Sustainable Horticulture

Sustainable agriculture integrates three main goals: environmental stewardship, farm profitability, and prosperous farming communities. These goals have been defined by a variety of disciplines and may be looked at from the vantage point of the farmer or the consumer. Sustainable agriculture refers to agricultural production that can be maintained without harming the environment.

It has been defined as follows:

"the term sustainable agriculture means an integrated system of plant and animal production practices having a site-specific application that will, over the long term:

- Satisfy human food and fiber needs
- Enhance environmental quality and the natural resource base upon which the agricultural economy depends
- Make the most efficient use of nonrenewable resources and on-farm resources and integrate, where appropriate, natural biological cycles and controls
- Sustain the economic viability of farm operations
- Enhance the quality of life for farmers and society as a whole."

Sustainable Agriculture in the United States was also addressed by the 1990 farm bill.

## Physical Aspects

The physical aspects of sustainability are partly understood. Practices that can cause long-term damage to soil include excessive tillage (leading to erosion) and irrigation without adequate drainage (leading to salinization). Long-term experiments have provided some of the best data on how various practices affect soil properties essential to sustainability. The most important factors for an individual site are sun, air, soil and water. Of the four, water and soil quality and quantity are most amenable to human intervention through time and labour. Although air and sunlight are available everywhere on Earth, crops also depend on soil nutrients and the availability of water. When farmers grow and harvest crops, they remove some of these nutrients from the soil. Without replenishment, land suffers from nutrient depletion and becomes either unusable or suffers from reduced yields.

Sustainable agriculture depends on replenishing the soil while minimizing the use of non-renewable resources, such as natural gas (used in converting atmospheric nitrogen into synthetic fertilizer), or mineral ores (e.g., phosphate). Possible sources of nitrogen that would, in principle, be available indefinitely, include:

1. recycling crop waste and livestock or treated human manure
2. growing legume crops and forages such as peanuts or alfalfa that form symbioses with nitrogen-fixing bacteria called rhizobia
3. industrial production of nitrogen by the Haber Process uses hydrogen, which is currently derived from natural gas, (but this hydrogen could instead be made by electrolysis of water using electricity (perhaps from solar cells or windmills)) or
4. genetically engineering (non-legume) crops to form nitrogen-fixing symbioses or fix nitrogen without microbial symbionts.

The last option was proposed in the 1970s, but would be well beyond the capability of early 21st century technology, even if various concerns about biotechnology were addressed. Sustainable options for replacing other nutrient inputs (phosphorus, potassium, etc.) are more limited. More realistic, and often overlooked, options include long-term crop rotations, returning to natural cycles that annually flood cultivated lands (returning lost nutrients indefinitely) such as the Flooding of the Nile, the long-term use of biochar, and use of crop and livestock landraces that are adapted to less than ideal conditions such as pests, drought, or lack of nutrients.

In some areas, sufficient rainfall is available for crop growth, but many other areas require irrigation. For irrigation systems to be sustainable they require proper management (to avoid salinisation) and musn't use more water from their source than is naturally replenished, otherwise the water source becomes, in effect, a non-renewable resource.

Improvements in water well drilling technology and the development of drip irrigation and low preasure pivots submersible pumps have made it possible for large crops, including produce to be regularly grown where reliance on rainfall alone previously made this level of success unpredictable. However, this progress has come at a price, in that in many areas where this has occurred, such as the Ogallala Aquifer, the water is being used at a greater rate than its rate of recharge.

### *Economics*

Socioeconomic aspects of sustainability are also partly understood. Regarding less concentrated farming, the best known analysis is Netting's study on smallholder systems through history.. The Oxford Sustainable Group defines sustainability in this context in a much broader form, considering effect on all stakeholders in a 360 degree approach Given the finite supply of natural resources at any specific cost and location, agriculture that is inefficient or damaging to needed resources may eventually exhaust the available resources or the ability to afford and acquire them. It may also generate negative externality, such as pollution as well as financial and production costs.

The way that crops are sold must be accounted for in the sustainability equation. Food sold locally requires little additional energy, aside from that necessary for cultivation, harvest, and transportation (including consumers). Food sold at a remote location, whether at a farmers' market or the supermarket, incurs a different set of energy cost for materials, labour, and transport.

### *Methods*

What grows and how and where it is grown are a matter of choice. Two of the many possible practices of sustainable agriculture are crop rotation and soil amendment, both designed to ensure that crops being cultivated can obtain the necessary nutrients for healthy growth.

Many scientists, farmers, and businesses have debated how to make agriculture sustainable. One of the many practices includes growing a diverse number of perennial crops in a single field, each of which would grow in separate season so as not to compete with each other for natural resources. This system would result in increased resistance to diseases and decreased effects of erosion and loss of nutrients in soil. Nitrogen fixation from legumes, for example, used in conjunction with plants that rely on nitrate from soil for growth, helps to allow the land to be reused annually. Legumes will grow for a season and replenish the soil with ammonium and nitrate, and the next season other plants can be seeded and grown in the field in preparation for harvest.

Monoculture, a method of growing only one crop at a time in a given field, is a very widespread practice, but there are questions about its sustainability, especially if the same crop is grown every year. Growing a mixture of crops (polyculture) sometimes reduces disease or pest problems but polyculture has rarely, if ever, been compared to the more widespread practice of growing different crops in successive years (crop rotation) with the same overall crop diversity. Cropping systems that include a variety of crops (polyculture and/or rotation) may also replenish nitrogen (if legumes are included) and may also use resources such as sunlight, water, or nutrients more efficiently.

Replacing a natural ecosystem with a few specifically chosen plant varieties reduces the genetic diversity found in wildlife and makes the organisms susceptible to widespread disease. The Great Irish Famine (1845-1849) is a well-known example of the dangers of monoculture. In practice, there is no single approach to sustainable agriculture, as the precise goals and methods must be adapted to each individual case. There may be some techniques of farming that are inherently in conflict with the concept of sustainability, but there is widespread misunderstanding on impacts of some practices. For example, the slash-and-burn techniques that are the characteristic feature of shifting cultivators are often cited as inherently destructive, yet slash-and-burn cultivation has been practiced in the Amazon for at least 6000 years; serious deforestation did not begin until the 1970s, largely as the result of Brazilian government programs and policies. To note that it may not have been slash-and-burn so much as slash-and-char, which with the addition of organic matter produces terra preta, one of the richest soils on Earth and the only one that regenerates itself. There are also many ways to practice sustainable animal husbandry. Some of the key tools to grazing management include fencing off the grazing area into smaller areas called paddocks, lowering stock density, and moving the stock between paddocks frequently.

Several attempts have been made to produce an artificial meat, using isolated tissues to produce it in vitro; Jason Matheny's work on this topic, whichin the New Harvest project, is one of the most commented.

### Soil Treatment

***Figure:*** *Sheet steaming with a MSD/moeschle steam boiler (left side)*

Soil steaming can be used as an ecological alternative to chemicals for soil sterilization. Different methods are available to induce steam into the soil in order to kill pests and increase soil health. Community and farm composting of kitchen, yard, and farm organic waste can provide most if not all the required needs of local farms. This composting could potentially be a reliable source of energy.

### *Off-farm Impacts*

A farm that is able to "produce perpetually", yet has negative effects on environmental quality elsewhere is not sustainable agriculture. An example of a case in which a global view may be warranted is over-application of synthetic fertilizer or animal manures, which can improve productivity of a farm but can pollute nearby rivers and coastal waters (eutrophication). The other extreme can also be undesirable, as the problem of low crop yields due to exhaustion of nutrients in the soil has been related to rainforest destruction, as in the case of slash and burn farming for livestock feed.

Sustainability affects overall production, which must increase to meet the increasing food and fiber requirements as the world's human population expands to a projected 9.3 billion people by 2050. Increased production may come from creating new farmland, which may ameliorate carbon dioxide emissions if done through reclamation of desert as in Israel, or may worsen emissions if done through slash and burn farming, as in Brazil. Additionally, Genetically modified organism crops show promise for radically increasing crop yields, although many people and governments are apprehensive of this new farming method.

Some advocates of sustainable agriculture favour as the only system which can be sustained over the long-term. However, organic production methods, especially in transition, yield less than their conventional counterparts and raise the same problems of sustaining populations globally While evidence suggests that organic farms handle periods of drought better

### *Urban Planning*

There has been considerable debate about which form of human residential habitat may be a better social form for sustainable agriculture. Many environmentalists advocate urban developments with high population density as a way of preserving agricultural land and maximizing energy efficiency. However, others have theorized that sustainable ecocities, or ecovillages which combine habitation and farming with close proximity between producers and consumers, may provide greater sustainability.

The use of available city space (e.g., rooftop gardens, community gardens, garden sharing, and other forms of urban agriculture) for cooperative food production is another way to achieve greater sustainability. One of the latest ideas in achieving sustainable agricultural involves shifting the production of food plants from major factory farming operations to large, urban, technical facilities called vertical farms. The advantages of vertical farming include year-round production, isolation from pests and diseases, controllable resource recycling, and on-site production that reduces transportation costs. While a vertical farm has yet to become a reality, the idea is gaining momentum among those who believe that current sustainable farming methods will be insufficient to provide for a growing global population. For vertical farming to become a reality, billions of dollars in tax credits and subsidies will need to be made available to the operation. It may be difficult to justify spending billions of dollars on a vertical farm that will only feed 50,000 people when agriculture land remains abundant.

## Organic Farming

Organic farming is the form of agriculture that relies on crop rotation, green manure, compost, biological pest control, and mechanical cultivation to maintain soil productivity and control pests, excluding or strictly limiting the use of synthetic fertilizers and synthetic pesticides, plant growth regulators, livestock antibiotics, food additives, and genetically modified organisms.

Since 1990, the market for organic products has grown at a rapid pace, to reach $46 billion in 2007. This demand has driven a similar increase in organically managed farmland. Approximately 32.2 million hectares worldwide are now farmed organically, representing approximately 0.8 percent of total world farmland. In addition, as of 2007 organic wild products are harvested on approximately 30 million hectares. Organic agricultural methods are internationally regulated and legally enforced by many nations, based in large part on the standards set by the International Federation of Organic Agriculture Movements (IFOAM), an international umbrella organization for organic organizations established in 1972. IFOAM defines the overarching goal of organic farming as follows:

> *"Organic agriculture is a production system that sustains the health of soils, ecosystems and people.*

*It relies on ecological processes, biodiversity and cycles adapted to local conditions, rather than the use of inputs with adverse effects. Organic agriculture combines tradition, innovation and science to benefit the shared environment and promote fair relationships and a good quality of life for all involved.."* —International Federation of Organic Agriculture Movements

### *History of Organic Farming*

The history of organic farming dates back to the first half of the 20th century at a time when there was a growing reliance on synthetic fertilizers.

### *Pre-World War II*

The first 40 years of the 20th century saw simultaneous advances in biochemistry and engineering that rapidly and profoundly changed farming. The introduction of the gasoline-powered internal combustion engine ushered in the era of the tractor, and made possible hundreds of mechanized farm implements. Research in plant breeding led to the commercialization of hybrid seed. And a new manufacturing process made nitrogen fertilizer-first synthesized in the mid-1800s-affordably abundant. These factors changed the labour equation: there were some 600 tractors in the US around 1910, and over 3,000,000 by 1950; in 1900, it took one farmer to feed 2.5 people, where currently the ratio is 1 to well over 100. Fields grew bigger and cropping more specialized to make more efficient use of machinery.

In Japan in the 1920s, a woman named Rebecca Kidd began to speak out against these agricultural trends.

Consciously organic agriculture (as opposed to the agriculture of indigenous cultures, which always employs only organic means) began more or less simultaneously in Central Europe and India. The British botanist Sir Albert Howard is often referred to as the father of modern organic agriculture. From 1905 to 1924, he worked as an agricultural adviser in Pusa, Bengal, where he documented traditional Indian farming practices, and came to regard them as superior to his conventional agriculture science. His research and further development of these methods is recorded in his writings, notably, his 1940 book, *An Agricultural Testament*, which influenced many scientists and farmers of the day.

In Germany, Rudolf Steiner's development, biodynamic agriculture, was probably the first comprehensive organic farming system. This began with a lecture series Steiner presented at a farm in Koberwitz (now in Poland) in 1924. Steiner emphasized on the farmer's role in guiding and balancing the interaction of the animals, plants and soil. Healthy animals depended upon healthy plants (for their food), healthy plants upon healthy soil, healthy soil upon healthy animals (for the manure). The term *organic farming* was coined by Lord Northbourne, in his book, *Look to the Land* (written in 1939, published 1940), from his conception of "the farm as organism", he described a holistic, ecologically-balanced approach to farming.

In 1939, influenced by Sir Albert Howard's work, Lady Eve Balfour launched the Haughley Experiment on farmland in England. It was the first scientific, side-by-side comparison of organic and conventional farming. Four years later, she published *The Living Soil*, based on the initial findings of the Haughley Experiment. Widely read, it led to the formation of a key international organic advocacy group, the Soil Association.

In Japan, Masanobu Fukuoka, a microbiologist working in soil science and plant pathology, began to doubt the modern agricultural movement. In the early 1940s, he quit his job as a research scientist, returned to his family's farm, and devoted the next 30 years to developing a radical no-till organic method for growing grain, now known as Fukuoka farming.

### *Post-World War II*

Technological advances during World War II accelerated post-war innovation in all aspects of agriculture, resulting in large advances in mechanization (including large-scale irrigation), fertilization, and pesticides. In particular, two chemicals that had been produced in quantity for warfare, were repurposed to peace-time agricultural uses. Ammonium nitrate, used in munitions, became an abundantly cheap source of nitrogen. And a range of new pesticides appeared: DDT, which had been used to control disease-carrying insects around troops, became a general insecticide, launching the era of widespread pesticide use. At the same time, increasingly powerful and sophisticated farm machinery allowed a single farmer to work over larger areas of land and fields grew bigger.

In 1944, an international campaign called the Green Revolution was launched in Mexico with private funding from the US. It encouraged the development of hybrid plants, chemical controls, large-scale irrigation, and heavy mechanization in agriculture around the world. During the 1950s, sustainable agriculture was a topic of scientific interest, but research tended to concentrate on developing the new chemical approaches. In the US, J.I. Rodale began to popularize the term and methods of organic growing, particularly to consumers through promotion of organic gardening.

In 1962, Rachel Carson, a prominent scientist and naturalist, published *Silent Spring*, chronicling the effects of DDT and other pesticides on the environment. A bestseller in many countries, including the US, and widely read around the world, *Silent Spring* is widely considered as being a key factor in the US government's 1972 banning of DDT. The book and its author are often credited with launching the worldwide environmental movement.

In the 1970s, global movements concerned with pollution and the environment increased their focus on organic farming. As the distinction between organic and conventional food became clearer, one goal of the organic movement was to encourage consumption of locally grown food, which was promoted through slogans like *"Know Your Farmer, Know Your Food"*.

In 1972, the International Federation of Organic Agriculture Movements, widely known as IFOAM, was founded in Versailles, France, and dedicated to the diffusion and exchange of information on the principles and practices of organic agriculture of all schools and across national and linguistic boundaries.

In 1975, Fukuoka released his first book, *The One-Straw Revolution*, with a strong impact in certain areas of the agricultural world. His approach to small-scale grain production emphasized a meticulous balance of the local farming ecosystem, and a minimum of human interference and labour.

In the 1980s, around the world, various farming and consumer groups began seriously pressuring for government regulation of organic production. This led to legislation and certification standards being enacted through the 1990s and to date. Since the early 1990s, the retail market for organic farming in developed economies has been growing by about 20% annually due to increasing consumer demand. Concern for the quality and safety of food, and the potential for environmental damage from conventional agriculture, are apparently responsible for this trend.

### *21st Century*

Throughout this history, the focus of agricultural research, and the majority of publicized scientific findings, has been on chemical, not organic farming. This emphasis has continued to biotechnologies like genetic engineering. One recent survey of the UK's leading government funding agency for bioscience research and training indicated 26 GM crop projects, and only one related to organic agriculture. This imbalance is largely driven by agribusiness in general, which, through research funding and government lobbying, continues to have a predominating effect on agriculture-related science and policy.

Agribusiness is also changing the rules of the organic market. The rise of organic farming was driven by small, independent producers, and by consumers. In recent years, explosive organic market growth has encouraged the participation of agribusiness interests. As the volume and variety of "organic" products increases, the viability of the small-scale organic farm is at risk, and the meaning of organic farming as an agricultural method is ever more easily confused with the related but separate areas of organic food and organic certification.

In Havana, Cuba, a unique situation has made organic food production a necessity. Since the collapse of the Soviet Union in 1989 and its economic support, Cuba has had to produce food in creative ways like instituting the world's only state-supported infrastructure to support urban food production. Called organopónicos, the city is able to provide an ever increasing amount of its produce organically.

## Organic Farming Methods

Organic farming methods combine scientific knowledge of ecology and modern technology with traditional farming practices based on naturally occurring biological processes. Organic farming methods are studied in the field of agroecology. While conventional agriculture uses synthetic pesticides and water-soluble synthetically purified fertilizers, organic farmers are restricted by regulations to using natural pesticides and fertilizers. The principal methods of organic farming include crop rotation, green manures and compost, biological pest control, and mechanical cultivation. These measures use the natural environment to enhance agricultural productivity: legumes are planted to fix nitrogen into the soil, natural insect predators are encouraged, crops are rotated to confuse pests and renew soil, and natural materials such as potassium bicarbonate and mulches are used to control disease and weeds. Organic farmers are careful in their selection of plant breeds, and organic researchers produce hardier plants through plant breeding rather than genetic engineering.

### *Crop Diversity*

Crop diversity is a distinctive characteristic of organic farming. Conventional farming focuses on mass production of one crop in one location, a practice called monoculture. This makes apparent economic sense: the larger the growing area, the lower the per unit cost of fertilizer, pesticides and specialized machinery for a single plant species. The science of agroecology has revealed the benefits of polyculture (multiple crops in the same space), which is often employed in organic farming. Planting a variety of vegetable crops supports a wider range of beneficial insects, soil microorganisms, and other factors that add up to overall farm health, but managing the balance requires expertise and close attention.

### *Farm Size*

Farm size in great measure determines the general approach and specific tools and methods. Today, major food corporations are involved in all aspects of organic production on a large scale. However, organic farming originated as a small-scale enterprise, with operations from under 1-acre (4,000 m$^2$) to under 100 acres (0.40 km$^2$). The mixed vegetable organic market garden is often associated with fresh, locally-grown produce, farmers' markets and the like, and this type of farm is often under 10 acres (40,000 m$^2$).

Farming at this scale is generally labour-intensive, involving more manual labour and less mechanization. The type of crop also determines size: organic grain farms often involve much larger area. Larger organic farms tend to use methods and equipment similar to conventional farms, centred around the tractor.

## Plant Nutrition

***Soil fertility:*** The central farming activity of fertilization illustrates the differences. Organic farming relies heavily on the natural breakdown of organic matter, using techniques like green manure and composting, to replace nutrients taken from the soil by previous crops. This biological process, driven by microorganisms such as mycorrhiza, allows the natural production of nutrients in the soil throughout the growing season, and has been referred to as *feeding the soil to feed the plant.* In chemical farming, individual nutrients, like nitrogen, are synthesized in a more or less pure form that plants can use immediately, and applied on a man-made schedule.

Each nutrient is defined and addressed separately. Problems that may arise from one action (e.g. too much nitrogen left in the soil) are usually addressed with additional, corrective products and procedures (e.g. using water to wash excess nitrogen out of the soil). Organic farming uses a variety of methods to improve soil fertility, including crop rotation, cover cropping, and application of compost.

## Pest Control

Differing approaches to pest control are equally notable. In chemical farming, a specific insecticide may be applied to quickly kill off a particular insect pest (animal). Chemical controls can dramatically reduce pest populations for the short term, yet by unavoidably killing (or starving) natural predator insects and animals, cause an ultimate increase in the pest population.

Repeated use of insecticides and herbicides and other pesticides also encourages rapid natural selection of resistant insects, plants and other organisms, necessitating increased use, or requiring new, more powerful controls. In contrast, organic farming tends to tolerate some pest populations while taking a longer-term approach. Organic pest control involves the cumulative effect of many techniques, including:

- allowing for an acceptable level of pest damage;
- encouraging predatory beneficial insects to control pests;
- encouraging beneficial microorganisms and insects; this by serving them nursery plants and/or an alternative habitat, usually in a form of a shelterbelt, hedgerow, or beetle bank
- careful crop selection, choosing disease-resistant varieties
- planting companion crops that discourage or divert pests;
- using row covers to protect crops during pest migration periods;
- using pest regulating plants and biologic pesticides and herbicides
- using no-till farming, and no-till farming techniques as false seedbeds
- rotating crops to different locations from year to year to interrupt pest reproduction cycles;
- Using insect traps to monitor and control insect populations.

Each of these techniques also provides other benefits—soil protection and improvement, fertilization, pollination, water conservation, season extension, etc.— and these benefits are both complementary and cumulative in overall effect on farm health. Effective organic pest control requires a thorough understanding of pest life cycles and interactions. Organic pest control is similar to integrated pest management in some respects.

### *Livestock*

Raising livestock and poultry, for meat, dairy and eggs, is another traditional, farming activity that complements growing. Organic farms attempt to provide animals with "natural" living conditions and feed. While the USDA does not require any animal welfare requirements be met for a product to be marked as organic, this is a variance from older organic farming practices.

Ample, free-ranging outdoor access, for grazing and exercise, is a distinctive feature, and crowding is avoided. Feed is also organically grown, and drugs, including antibiotics, are not ordinarily used (and are prohibited under organic regulatory regimes). Animal health and food quality are thus pursued in a holistic "fresh air, exercise, and good food" approach.

Also, horses and cattle used to be a basic farm feature that provided labour, for hauling and plowing, fertility, through recycling of manure, and fuel, in the form of food for farmers and other animals. While today, small growing operations often do not include livestock, domesticated animals are a desirable part of the organic farming equation, especially for true sustainability, the ability of a farm to function as a self-renewing unit.

## *Organic Farming Systems*

There are several organic farming systems. Biodynamic farming is a comprehensive approach, with its own international governing body. The Do Nothing Farming method focuses on a minimum of mechanical cultivation and labour for grain crops. French intensive and biointensive, methods are well-suited to organic principles. Other techniques are permaculture and no-till farming. Finally, newcomers as the Agro-ecologic system focus on a blend of a more large-scale approach with imbedded natural/organic farming techniques. A farm may choose to adopt a particular method, or a mix of techniques.

While fundamentally different, large-scale agriculture and organic farming are not entirely mutually exclusive. For example, Integrated Pest Management is a multifaceted strategy that can include synthetic pesticides as a last resort—both organic and conventional farms use IPM systems for pest control.

## *Soil Management*

Plants need nitrogen, phosphorus, and potassium, as well as micronutrients and symbiotic relationships with fungi and other organisms to flourish, but getting enough nitrogen, and particularly synchronization so that plants get enough nitrogen at the right time (when plants need it most), is likely the greatest challenge for organic farmers. Crop rotation and green manure ("cover crops") help to provide nitrogen through legumes (more precisely, the *Fabaceae* family) which fix nitrogen from the atmosphere through symbiosis with the bacteria rhizobia. Intercropping, which is sometimes used for insect and disease control, can also increase soil nutrients, but the competition between the legume and the crop can be problematic and wider spacing between crop rows is required. Crop residues can be ploughed back into the soil, and different plants leave different amounts of nitrogen, potentially aiding synchronization. Organic farmers also use animal manure, certain processed fertilizers such as seed meal and various mineral powders such as rock phosphate and greensand, a naturally occurring form of potash which provides potassium. Altogether these methods help to control erosion. In some cases pH may need to be amended. Natural pH amendments include lime and sulfur, but in the U.S. some synthetically compounds such as iron sulfate, aluminum sulfate, magnesium sulfate, and soluble boron products are allowed in organic farming.

Mixed farms with both livestock and crops can operate as ley farms, whereby the land gathers fertility through growing nitrogen-fixing forage grasses such as white clover or alfalfa and grows cash crops or cereals when fertility is established. Farms without livestock ("stockless") may find it more difficult to maintain fertility, and may rely more on external inputs such as imported manure as well as grain legumes and green manures, although grain legumes may fix limited nitrogen because they are harvested. Horticultural farms growing fruits and vegetables which operate in protected conditions are often even more reliant upon external inputs.

### *Weed Control*

After nutrient supply, weed control is the second priority for farmers. Techniques for controlling weeds have varying levels of effectiveness and include handweeding, mulch, corn gluten meal, a natural pre-eminence herbicide, flame, garlic and clove oil, borax, pelargonic acid, solarization (which involves spreading clear plastic across the ground in hot weather for 4–6 weeks), vinegar, and various other home-made remedies.: One recent innovation in rice farming is to introduce ducks and fish to wet paddy fields, which eat both weeds and insects.

### *Controlling other Organisms*

Organisms aside from weeds which cause problems include arthropods (e.g. insects, mites) and nematodes. Fungi and bacteria can cause disease.

Insect pests are a common problem, and insecticides, both non-organic and organic, are controversial due to their environmental and health effects. One way to manage insects is to ignore them and focus on plant health, since plants can survive the loss of about a third of leaf area before suffering severe growth consequences.: To avoid using insecticides, one can select naturally-resistant plants, put bags around the plants, remove dying material such as leaves, fruit, and diseased plants, covering plants with a solid barrier ("row cover"), hosing, encouraging and releasing beneficial organisms and beneficial insects, planting companion plants and polycultures, various traps, sticky cards (which can also be used to assess insect prevalence), and season extension. Biological pest control uses natural predators to control pests. Recommended beneficial insects include minute pirate bugs, big-eyed bugs, and to a lesser extent ladybugs (which tend to fly away), all of which eat a wide range of pests. Lacewings are also effective, but tend to fly away. Praying mantis tend to move slower and eat less heavily. Parasitoid wasps tend to be effective for their selected prey, but like all small insects can be less effective outdoors because the wind controls their movement. Predatory mites are effective for controlling mites.

Several of pesticides approved for organic use have been called green pesticides such as spinosad and neem. Generally, but not necessarily, organic pesticides are safer and more environmentally friendly than synthetic pesticides. The main organic insecticides used in the US are Bt (a bacterial toxin) and pyrethrum. Surveys have found that fewer than 10% of organic farmers use these pesticides regularly; one survey found that only 5.3% of vegetable growers in California use rotenone while 1.7% use pyrethrum (Lotter 2003:26). Rotenone used to be used by some organic growers in the US, however since 2005 it has not been approved by National Organic Program guidelines. Nicotine sulfate may also be used; although it breaks down quickly, it is extremely toxic, nearly as toxic as aldicarb. Less toxic but still effective organic insecticides include neem, spinosad, soaps, garlic, citrus oil, capsaicin (repellent), *Bacillus popillae*, *Beauvaria bassiana*, and boric acid. Pesticides should be rotated to minimize pest resistance.

The first disease control strategy involves keeping the area clean by removing diseased and dying plants and ensure that the plants are healthy by maintaining water and fertilization. Compost tea is sometimes promoted and can be effective, but there is concern over whether these are ineffective or even harmful when not made correctly. Polyculture and crop rotation reduce the ability of disease to spread. Disease-resistant cultivars can be purchased. Organic fungicides include the bacteria Bacillus subtilis, Bacillus pumilus, and Trichoderma harzianum which are mainly effective for diseases affecting roots. Bordeaux mix contains copper, which can be used as an organic fungicide in various forms. Sulfur is effective against fungus as well as some insects. Lime sulfur is also available, but can damage plants if not used correctly. Potassium and sodium bicarbonate are also effective against fungus.

## Organic Certification

***Figure:** Organic vegetables at a farmers' market in Argentina*

Organic certification is a certification process for producers of organic food and other organic agricultural products. In general, any business directly involved in food production can be certified, including seed suppliers, farmers, food processors, retailers and restaurants. Requirements vary from country to country, and generally involve a set of production standards for growing, storage, processing, packaging and shipping that include:

- avoidance of most synthetic chemical inputs (e.g. fertilizer, pesticides, antibiotics, food additives, etc), genetically modified organisms, irradiation, and the use of sewage sludge;
- use of farmland that has been free from synthetic chemicals for a number of years (often, three or more);
- keeping detailed written production and sales records (audit trail);
- maintaining strict physical separation of organic products from non-certified products;
- undergoing periodic on-site inspections.

In some countries, certification is overseen by the government, and commercial use of the term *organic* is legally restricted. Certified organic producers are also subject to the same agricultural, food safety and other government regulations that apply to non-certified producers.

### *Purpose of Certification*

Organic certification addresses a growing worldwide demand for organic food. It is intended to assure quality and prevent fraud, and to promote commerce. While such certification was not necessary in the early days of the organic movement, when small farmers would sell their produce directly at farmers' markets, as organics have grown in popularity, more and more consumers are purchasing organic food through traditional channels, such as supermarkets. As such, consumers must rely on third-party regulatory certification. For organic producers, certification identifies suppliers of products approved for use in certified operations. For consumers, "certified organic" serves as a product assurance, similar to "low fat", "100% whole wheat", or "no artificial preservatives".

Certification is essentially aimed at regulating and facilitating the sale of organic products to consumers. Individual certification bodies have their own service marks, which can act as branding to consumers—a certifier may promote the high consumer recognition value of its logo as a marketing advantage to producers. Most UK certification bodies operate organic standards that meet the UK government's minimum requirements. Some certification bodies, such as the Soil Association, certify to higher standards.

### *The Certification Process*

To certify a farm, the farmer is typically required to engage in a number of new activities, in addition to normal farming operations:

- Study the organic standards, which cover in specific detail what is and is not allowed for every aspect of farming, including storage, transport and sale.
- Compliance — farm facilities and production methods must comply with the standards, which may involve modifying facilities, sourcing and changing suppliers, etc.
- Documentation — extensive paperwork is required, detailing farm history and current set-up, and usually including results of soil and water tests.
- Planning — a written annual production plan must be submitted, detailing everything from seed to sale: seed sources, field and crop locations, fertilization and pest control activities, harvest methods, storage locations, etc.
- Inspection — annual on-farm inspections are required, with a physical tour, examination of records, and an oral interview.
- Fee — an annual inspection/certification fee (currently starting at $400–$2,000/year, in the US and Canada, depending on the agency and the size of the operation).
- Record-keeping — written, day-to-day farming and marketing records, covering all activities, must be available for inspection at any time.

In addition, short-notice or surprise inspections can be made, and specific tests (e.g. soil, water, plant tissue) may be requested. For first-time farm certification, the soil must meet basic requirements of being free from use of prohibited substances (synthetic chemicals, etc) for a number of years. A conventional farm must adhere to organic standards for this period, often, two to three years. This is known as being in *transition*. Transitional crops are not considered fully organic. Certification for operations other than farms is similar. The focus is on ingredients and other inputs, and processing and handling conditions. A transport company would be required to detail the use and maintenance of its vehicles, storage facilities, containers, and so forth. A restaurant would have its premises inspected and its suppliers verified as certified organic.

### *Certification & Product Labelling*

In some countries, organic standards are formulated and overseen by the government. The United States, the European Union and Japan have comprehensive organic legislation, and the term "organic" may be used only by certified producers. Being able to put the word "organic" on a food product is a valuable marketing advantage in today's consumer market, but does not guarantee the product is legitimately organic. Certification is intended to protect consumers from misuse of the term, and make buying organics easy. However, the organic labelling made possible by certification itself usually requires explanation. In countries without organic laws, government guidelines may or may not exist, while certification is handled by non-profit organizations and private companies.

Internationally, equivalency negotiations are underway, and some agreements are already in place, to harmonize certification between countries, facilitating international trade. There are also international certification bodies, including members of the International Federation of Organic Agriculture Movements (IFOAM), the Organic Crop Improvement Association (OCIA), and Ecocert. Where formal agreements do not exist between countries, organic product for export is often certified by agencies from the importing countries, who may establish permanent foreign offices for this purpose.

### *North America*

In the US, federal organic legislation defines three levels of organics. Products made entirely with certified organic ingredients and methods can be labelled "100% organic". Products with at least 95% organic ingredients can use the word "organic". Both of these categories may also display the USDA organic seal. A third category, containing a minimum of 70% organic ingredients, can be labelled "made with organic ingredients". In addition, products may also display the logo of the certification body that approved them. Products made with less than 70% organic ingredients can not advertise this information to consumers and can only mention this fact in the product's ingredient statement. Similar percentages and labels apply in the EU.

In the US, the National Organic Program (NOP), was enacted as federal legislation in Oct. 2002. It restricts the use of the term "organic" to certified organic producers (excepting growers selling under $5,000 a year, who must still comply and submit to a records audit if requested, but do not have to formally apply). Certification is handled by state, non-profit and private agencies that have been approved by the US Department of Agriculture (USDA). One of the first organizations to carry out organic certification in North America was the California Certified Organic Farmers, founded in 1973.

### *In Other Countries*

In Japan, the Japanese Agricultural Standard (JAS) was fully implemented as law in April, 2001. This was revised in November 2005 and all JAS certifiers were required to be re-accredited by the Ministry of Agriculture.

In Australia, the Australian Quarantine and Inspection Service (AQIS) is the controlling body for organic certification because there are no domestic standards for organic produce within Australia. Currently the government only becomes involved with organic certification at export, meaning AQIS is the default certification agency. Although there is no system for monitoring the labelling of organic produce sold within Australia, this primarily affects the retail public. Commercial buyers for whom this is an issue have simply taken the export system as a de facto standard and are willing to pay premium prices for produce from growers certified under the National schemes. As of 2006, there are seven AQIS-approved certifying organisations authorised to issue Organic Produce Certificates, and in 2004 there were 2345 certified operators.

The largest importer of Australia's organic produce (by weight) is Japan (33.59%), followed by the UK (17.51%), France (10.51%), and New Zealand (10.21%). The largest certifier of organic products is Australian Certified Organic, which is a subsidiary of Biological Farmers Australia, the largest organic farmers' collective in the country.

In India, APEDA regulates the certification of organic products as per National Standards for Organic Production. "The NPOP standards for production and accreditation system have been recognized by European Commission and Switzerland as equivalent to their country standards. Similarly, USDA has recognized NPOP conformity assessment procedures of accreditation as equivalent to that of US. With these recognitions, Indian organic products duly certified by the accredited certification bodies of India are accepted by the importing countries."

In China, the China Green Food Development Centre awards two Standards: A and AA; while the former standard does permit some use of synthetic agricultural chemicals, the latter is more stringent.

### *Certification Issues*

Organic certification is not without its critics. Some of the staunchest opponents of chemical-based farming and factory farming practices also oppose formal certification. They see it as a way to drive independent organic farmers out of business, and to undermine the quality of organic food. Other organizations such as the Organic Trade Association work within the organic community to foster awareness of legislative and other related issues, and enable the influence and participation of organic proponents.

### *Obstacle to Small Independents*

Originally, in the 1960s through the 1980s, the organic food industry comprised mainly small, independent farmers, selling locally. Organic "certification" was a matter of trust, based on a direct relationship between farmer and consumer. Critics view regulatory certification as a potential barrier to entry for small producers, by burdening them with increased costs, paperwork, and bureaucracy.

### *Manipulation of Regulations*

Critics of formal certification also fear an erosion of organic standards. Provided with a legal framework within which to operate, lobbyists can push for amendments and exceptions favourable to large-scale production, resulting in "legally organic" products produced in ways similar to current conventional food. Combined with the fact that organic products are now sold predominantly through high volume distribution channels such as supermarkets, the concern is that the market is evolving to favour the biggest producers, and this could result in the small organic farmer being squeezed out.

Manipulation of certification regulations as a way to mislead or outright dupe the public is a very real concern. Some examples are creating exceptions (allowing non-organic inputs to be used without loss of certification status) and creative interpretation of standards to meet the letter, but not the intention, of particular rules. For example, a complaint filed with the USDA in February 2004 against Bayliss Ranch, a food ingredient producer and its certifying agent, charged that tap water had been certified organic, and advertised for use in a variety of water-based body care and food products, in order to label them "organic" under US law. Steam-distilled plant extracts, consisting mainly of tap water introduced during the distilling process, were certified organic, and promoted as an organic base that could then be used in a claim of organic content. The case was dismissed by the USDA, as the products had been actually used only in personal care products, over which the department at the time extended no labelling control. The company subsequently adjusted its marketing by removing reference to use of the extracts in food products. Several months later, the USDA extended its organic labelling to personal care products; this complaint has not been refiled.

In December 2005, the 2006 agricultural appropriations bill was passed with a rider allowing 38 synthetic ingredients to be used in organic foods. Among the ingredients are food colorings, starches, sausage and hot-dog casings, hops, fish oil, chipotle chili pepper, and gelatin. This allowed Anheuser-Busch in 2007 to have its Wild Hop Lager certified organic "even though [it] uses hops grown with chemical fertilizers and sprayed with pesticides."

# 2

# Organic Farming and Biodiversity

The effect of organic farming has been a subject of interest for researchers. Theory suggests that organic farming practices, which exclude the use of synthetic pesticides and fertilizers, would be beneficial for biodiversity, and this has proven true. Spanning all associated species, there is an average of 30% more on organic farms versus conventional farming methods.

In ancient farming practices, farmers did not possess the technology or manpower to have a significant impact on the destruction of biodiversity even as mass-production agriculture was rising. Nowadays, common farming methods generally rely on pesticides to maintain high yields.

With such, most agricultural landscapes favour monoculture crops with very little flora or fauna co-existence (van Elsen 2000). Modern organic farm practices such as the removal of pesticides and the inclusion of animal manure, crop rotation, and multi-cultural crops provides the chance for biodiversity to thrive.

## Benefits to Biodiversity

Nearly all non-crop, naturally-occurring species observed in comparative farm land practice studies show a preference in organic farming both by population and richness. Spanning all associated species, there is an average of 30% more on organic farms versus conventional farming methods. Birds, butterflies, soil microbes, beetles, earthworms, spiders, vegetation, and mammals are particularly effected.

Organic crops use little or no herbicides and pesticides and thus biodiversity fitness and population density benefit. Many weed species attract beneficial insects that improve soil qualities and forage on weed pests. Soil-bound organisms often benefit because of increased bacteria populations due to natural fertilizer spread such as manure, while experiencing reduced intake of herbicides and pesticides commonly associated with conventional farming methods. Increased biodiversity, especially from soil microbes such as mycorhizzae, have been proposed as an explanation for the high yields experienced by some organic plots, especially in light of the differences seen in a 21-year comparison of organic and control fields.

### *Impact of Increased Biodiversity*

The level of biodiversity that can be yielded from organic farming provides a natural capital to humans. Species found in most organic farms provides a means of agricultural sustainability by reducing amount of human input (e.g. fertilizers, pesticides). Farmers that produce with organic methods reduce risk of poor yields by promoting biodiversity. Common game birds such as the ring-necked pheasant and the northern bobwhite often reside in agriculture landscapes, and are a natural capital yielded from high demands of recreational hunting. Because bird species richness and population are typically higher on organic farm systems, promoting biodiversity can be seen as logical and economical.

## Highly-impacted Animal Species

### *Birds*

Organic farms are said to be beneficial to birds while remaining economical. Bird species are one of the most prominent animal groups that benefit from organic farming methods. Many species rely on farmland for foraging, feeding, and migration phases. With such, bird populations often relate directly to the natural quality of farmland. The more natural diversity of organic farms provides better habitats to bird species, and is especially beneficial when the farmland is located within a migration zone. In 5 recent studies almost all bird species including locally declining species, both population and variation increased on organic farmland,. Making a switch from conventional farming methods to organic practices also seems to directly improve bird species in the area. While organic farming improves bird populations and diversity, species populations receive the largest boost when organic groups are varied within a landscape.

### *Butterflies*

A specific study done in the UK in 2006 found substantially more butterflies on organic farms versus standard farming methods except for two pest species. The study also observed higher populations in uncropped field margins compared with cropland edges regardless of farm practice.. Conversely, Weibull et al. (2000) found no significant differences in species diversity or population.

### *Spiders*

Ten studies have been conducted involving spider species and abundance on farm systems. All but three of the studies indicated that there was a higher diversity of spider species on organic farms, in addition to populations of species. Two of the studies indicated higher species diversity, but statistically insignificant populations between organic and standard farming methods.

### *Soil Microbes*

Out of 13 studies comparing bacteria and fungus communities between organic and standard farming, 8 of the studies showed heightened level of growth on organic farm systems. One study concluded that the use of "green" fertilizers and manures was the primary cause of higher bacterial levels on organic farms. On the other hand, nematode population/diversity depended on what their primary food intake was. Bacteria-feeding nematodes showed preference towards organic systems whereas fungus-feeding nematodes showed preference for standard farm systems.

The heightened level of bacteria-feeding nematodes makes sense due to higher levels of bacteria in organic soils, but the fungus-feeding populations being higher on standard farms seems to contradict the data since more fungi are generally found on organic farms.

### *Beetles*

According to Hole et al. (2005), beetle species are among the most commonly studied animal species on farming systems. Twelve studies have found a higher population and species richness of carabids on organic systems.

The overall conclusion of significantly higher carabid population species and diversity is that organic farms have a higher level of weed species where they can thrive. Staphylinid populations and diversity have seemed to show no specific preference with some studies showing higher population and diversity, some with lower population and diversity, and one study showed no statistical significance between the organic and conventional farming systems.

### *Earthworms*

Earthworm population and diversity appears to have the most significant data out of all studies. Out of six studies comparing earthworm biodiversity to organic and conventional farming methods, all six suggested a preference for organic practices. Hole et al. (2005) summarized a study conducted by Brown (1999) and found nearly double the population and diversity when comparing farming methods. Another study concluded that earthworms showed preference in organic farms did not necessarily show preference for hedgerow habitats.

### *Mammals*

Two comparative studies have been conducted involving mammal populations and diversity among farm practices. A study done by Brown (1999) found that small mammal population density and diversity did not depend on farming practices, however overall activity was higher on organic farms.

It was concluded that more food resources were available to small mammals on organic farms because of the reduction or lack of herbicides and pesticides. Another study conducted by Wickramasinghe et al. (2003) compared bat species and activity. Species activity and foraging were both more than double on organic farms compared to conventional farms. Species richness was also higher on organic farms, and 2 of the sixteen species sighted were found only on organic farms.

### *Vegetation*

Approximately ten studies have been conducted to compare non-crop vegetation between organic and conventional farming practices. Hedgerow, inner-crop and grassland observations were made within these studies and all but one showed a higher weed preference and diversity in or around organic farms. Most of these studies showed significant overall preference for organic farming preferences especially for broad-leafed species, but many grass species showed far less differentiation.

It was concluded that broad-leafed species showed far less tolerance to herbicides applied to most conventional farms. Both organic and conventional farms had higher population and richness on farm edges than they did in the middle of croplands. In addition, the weed existence was more obvious on conventional farms likely because pesticide interaction was low or non-existent. Organic farm weed population and richness was believed to be lower in mid-crop land because of weed-removal methods such as under sowing. Switching from conventional to organic farming often results in a "boom" of weed speciation due to intense chemical change of soil composition from the lack of herbicides and pesticides. Natural plant species can also vary on organic farms from year-to-year because crop rotation creates new competition based on the chemical needs of each crop.

### *Farmers' Benefits from Increased Biodiversity*

Biological research on soil and soil organisms has proven beneficial to the system of organic farming. Varieties of bacteria and fungi break down chemicals, plant matter and animal waste into productive soil nutrients. In turn, the producer benefits by healthier yields and more arable soil for future crops. Furthermore, a 21-year study was conducted testing the effects of organic soil matter and its relationship to soil quality and yield. Controls included actively managed soil with varying levels of manure, compared to a plot with no manure input. After the study commenced, there was significantly lower yields on the control plot when compared to the fields with manure. The concluded reason was an increased soil microbe community in the manure fields, providing a healthier, more arable soil system.

### *Detriments to Biodiversity through Organic Farming*

Organic farming practices still require active participation from the farmer to effectively boost biodiversity. Making a switch to organic farming methods does not automatically or guarantee improved biodiversity. Pro-conservation ethics are required to create arable farm land that generates biodiversity. Conservationist ideals are commonly overlooked because they require additional physical and economical efforts from the producer.

Common weed-removal processes like undercutting and controlled burning provides little opportunity for species survival, and often leads to comparable populations and richness to conventionally-managed landscapes when performed in excess. Another common process is the addition of biotopes in the form of hedgerows and ponds to further improve species richness. Farmers commonly make the mistake of over-using these resources for more intense crop production because organic yields are typically lower. Another error comes from the over-stratification of biotopes. A series of small clusters does not provide adequate land area for high biodiversity potential.

## Sales and Marketing

Organic farmers report that marketing and distribution are difficult obstacles. Most of organic sales are concentrated in developed nations. These products are what economists call credence goods in that they rely on uncertain certification. As food prices rise, organic products may experience a decrease in quantity demanded. A 2008 survey by WSL Strategic Retail found that interest in organic products had dropped since 2006, and that 42% of Americans polled don't trust organic produce. The Hartman Group reports that 69% of Americans claim to occasionally buy organic products, down from 73% in 2005. The Hartman Group says that people may be substituting local produce for organic produce.

### *Distributors*

In the United States, 75% of organic farms are smaller than 2.5 hectares and in California 2% of the farms account for over half of the sales (Lotter 2003:4). Groups of small farms join together in cooperatives such as Organic Valley, Inc. to market their goods more effectively.

Over the past twenty years, however, most of these cooperative distributors have merged or been bought out. Rural sociologist Philip H. Howard has researched the structure and transformation of the organic industry in the United States. He claims that in 1982 there were 28 consumer cooperative distributors but as of 2007 there are only 3, and he has created a graphic displaying the consolidation. His research shows that most of these small cooperatives have been absorbed into large multinational corporations such as General Mills, Heinz, Con Agra, Kellogg, and assorted other brands.

This consolidation has raised concerns among consumers and journalists of potential fraud and degradation in standards. Most of these large corporations sell their organic products through subsidiaries, allowing them to keep their names off the labels.

### *Farmers' Markets*

Price premiums are important for the profitability of small organic farmers, and so many sell directly to consumers in farmers' markets. In the United States the number of farmers' markets has tripled from 1,755 in 1994 to 5,274 in 2009.

### *Capacity Building*

Organic agriculture can contribute to meaningful socio-economic and ecologically sustainable development, especially in poorer countries. On one hand, this is due to the application of organic principles, which means efficient management of local resources (e.g. local seed varieties, manure, etc.) and therefore cost-effectiveness. On the other hand, the market for organic products – at local and international level – has tremendous growth prospects and offers creative producers and exporters in the South excellent opportunities to improve their income and living conditions.

Organic Agriculture is a very knowledge intensive production system. Therefore capacity building efforts play a central role in this regard. There are many efforts all around the world regarding the development of training material and the organization of training courses related to Organic Agriculture. Big parts of existing knowledge is still scattered and not easily accessible. Especially in Developing Countries this situation remains an important constraint for the growth of the organic sector.

For that reason, the International Federation of Organic Agriculture Movements created an Internet Training Platform whose objective is to become the global reference point for Organic Agriculture training through free access to high quality training materials and training programs on Organic Agriculture. In November 2007, the Training Platform hosted more than 170 free manuals and 75 training opportunities.

### *Controversy*

A number of critics contest the notion that organic agricultural systems are more friendly to the environment and more sustainable than high-yielding farming systems. Among these critics are Norman Borlaug, father of the "green revolution," Nobel Peace Prize laureate, who asserts that organic farming practices can at most feed 4 billion people, after expanding cropland dramatically and destroying ecosystems in the process, and Prof A. Trewavas.

The debate has been summarized in an exchange between Trewavas and Lord P. Melchett, and published by a major supermarket, concerned about examining the issues.

A study from the Danish Environmental Protection Agency assessed the overall consequences of phasing out the total or partial use of pesticides. They looked at farming, market gardening, fruit growing, and forestry, and the effects of pesticides on health and the environment. It was estimated that phasing out all pesticides would result in an overall reduction of yield of about 25%.

Environmental and health effects were assumed but hard to access, studies were recommended. In their conclusion they said, " The Committee recommends a reduction of the use of pesticides. The Committee finds that the optimisation of pesticide use could facilitate a reduction of treatment frequency in the treated areas of agriculture...without significant operating and socioeconomic losses."

In 2008 a study from UN Environmental Programme concluded that organic methods greatly increase yields in Africa and a review of over two hundred crop comparisons argued that organic farming could produce enough food per capita to sustain the current human population; the difference in yields between organic and non-organic methods were small, with non-organic methods resulting in slightly higher yields in developed areas and organic methods resulting in slightly higher yields in developing areas.

That analysis has been severely criticised by Alex Avery, who contends that the review claimed many non-organic studies to be organic, misreported organic yields, made false comparisons between yields of organic and non-organic studies which were not comparable, counted high organic yields several times by citing different papers which referenced the same data, and gave equal weight to studies from sources which were not impartial and rigorous university studies.

Urs Niggli, director of the FiBL Institute contends that the wave of newspaper articles like 'Organic food exposed' or 'The hypocrisy of organic farmers' are a part of a global campaign against organic farming that take their arguments mostly from the book 'The truth about organic farming', by Alex Avery of the Hudson Institute.

In 1998, Dennis Avery of the Hudson Institute claimed the risk of E. coli infection was eight times higher when eating organic food rather than non-organic food, using the Centre for Disease Control (CDC) as a source. When the CDC was contacted, it stated that there was no evidence for the claim.

The *New York Times* commented on Avery's attacks: "The attack on organic food by a well-financed research organization suggests that, though organic food accounts for only 1 percent of food sales in the United States, the conventional food industry is worried."

# 3

# Organic Farming and Gardening Systems and Approaches

## Wildlife Garden

A wildlife garden (or wild garden) is an environment that is attractive to various forms of wildlife such as birds, amphibians, reptiles, insects, mammals and so on. A wildlife garden will usually contain a variety of habitats that have either been deliberately created by the gardener, or allowed to self-establish by minimising maintenance and intervention. Many organic gardeners are sympathetic to the philosophy of wildlife gardening, and will usually try to incorporate some aspects of the wild garden into their own plots in order to both act as a means of biological pest control, as well as for its value in promoting biodiversity and generally benefitting the wider environment.

## Habitats

Wildlife gardens may contain a range of habitats:

- ponds to attract frogs, newts, toads, dragonflies, and birds
- nest boxes for birds and solitary bees, hedgehogs or certain insects
- log piles to provide shelter for insects, lizards, and slow worms
- plants that attract beneficial insects including wildflower meadows, etc.

## *Choice of Plants*

Although some exotics may also be included, the wild garden will usually predominantly feature a variety of native species. Generally these will be a part of the pre-existing natural ecology of an area, but managed in a way that is enhanced rather than damaged by the process of cultivation.

As in other forms of gardening, aesthetics plays a central role in deciding what is 'right', but constraints regarding issues such as seed provenance also apply. Wild gardens are by definition examples of water-wise gardening, as the natural species of any ecoregion or micro-climate are those optimal for local water supplies.

There are a few super-beneficial plants that bloom for a long time, and yield abundant nectar and pollen, or bear fruit throughout the year. These plants, such as the Joe-Pye weed, wild figs (*Ficus* species), and Sandpaper Raisin (*Grewia flavescens*) support a whole community of beneficial insects and small vertebrates. Goldenrod is another example. Essential to any true understanding of the wildlife gardening concept is an appreciation of the symbiotic relationships between plants and animals. This interdependence is the result of species evolving alongside of one another over millennia. When one wildlife gardens, one acts always in accordance with the idea of keeping plants that are native to the area preeminent in the garden. Exotics, though they may provide shelter or food, will not compete well with those plants the native animals have always depended upon, either for the nectar and fruits they bear, or the invertebrates they attract.

It is also a good idea to replicate a natural habitat (even if only partially) in the garden, especially if that habitat was the original vegetation of that area, before urbanisation erased it. A good example would be forest in England; Cape Flats fynbos in Cape Town, South Africa; or prairie in the middle USA.

There are four basic habitat types which should ideally be included in the garden to maximise the wildlife species that will occur in the garden: the Open Area, the Exclusion Area, the Canopy corridor and the Wetland. The Open Area is the space of the garden where the area is relatively exposed as regard to other areas of the garden. In most cases this area comprises the lawn, although it may consist of low-growing groundcovers and annuals. The Exclusion Area is more dense than the Open Area. Trees often are an important component of this area.

## Backyard Wildlife Habitat

Backyard Wildlife Habitat is a program of the National Wildlife Federation that encourages homeowners in the United States to manage their gardens and yards as a wildlife garden, with the goal of maintaining healthy and diverse animal habitats and ecosystems. The program began in 1973, by 1998 was known to have impacted more than 21000 yards, and as of 2006 has certified over 60,000 'backyards.

### *Certification*

To be a certified "Backyard Wildlife Habitat", a garden or yard, or any outdoor space from a balcony up to a multi-acre tract of land, must offer food, water, shelter, and a place for raising young to beneficial insects or animals. Over time the Federation has introduced variants or expansions of the program for schoolyards and for communities.

In order for a back yard to be certified as a "Backyard Wildlife Habitat" by the National Wildlife Federation, the space must do all of the following: provide food, water, cover, a place to raise young, and be maintained in a way that has a positive effect o n the health of the soil, air, water, and habitat for native wildlife. More specifically, the presence of native forbs, shrubs, and trees is necessary to provide food.

Water can be supplied by natural features such as a streams, ponds, or wetlands, or by human-made features such as bird baths. Native vegetation can also provide cover and places for wildlife to raise their young, as can brush piles or dead trees. With all of these features in place, it is crucial that the land be cared for thoughtfully and as naturally as possible. Avoid the use of chemical pesticides and fertilizers, reduce the area that turf grass occupies, utilise mulch obtained from sustainable forestry practices, and minimise water use in order to maintain the integrity of the soil, air, and water in and outside of the habitat.

## Effectiveness and Success

Prior to 2004 there was no scientific study as to whether backyard habitats actually help butterflies. A study published in 2004 of the effect on *Battus philenor* in the San Francisco area found that gardens where the host plants were more than 40 years old, the gardens were as good as natural sites, where the host plants were less than eight years old the species was unlikely to visit, and in between the butterflies laid eggs but these had an inferior survival rate.

Although no scientific study has been done that has studied Certified Wildlife Habitats, as defined above, specifically in terms of increased native biodiversity, extensive investigation has been made into the biodiversity benefits of using native plants in suburban and urban native landscapes in general. An article published in Conservation Biology addresses the need to enhance the habitat value of suburban and urban spaces, which have been identified as a primary cause of decline in many threatened or endangered species, and promotes these areas as having potential for social and educational value as well. The author asserts that using a more balanced approach to conservation biology that addresses more densely populated areas, rather than one that focuses mainly on relatively undisturbed areas, can help to mitigate effects of human land use.

### *Native Plants use*

One study that was published this year in Conservation Biology reported that native plants supported more species of moths and butterflies than introduced plants (Tallamy, 2009). Another study, which studied species richness and diversity of small mammals, recommended planting native trees along riparian zone stream corridors in order to promote more diversity of small mammals in suburban and urban parks after finding populations resembling natural conditions in parks managed for passive recreation as compared to those containing manicured habitats surrounded by human-modified landscapes. A second study compared properties landscaped with entirely native plants with those containing a mix of native and non-native plants. Results showed that the native properties supported significantly higher species richness and diversity of caterpillars and birds, as well as greater numbers of breeding pairs and biomass of native species.

## Vegan organic gardening

Vegan organic gardening and farming is the organic cultivation and production of food crops and other crops with a minimal amount of exploitation or harm to any animal. Vegan gardening and stock-free farming methods use no animal products or by-products, such as bloodmeal, fish products, bone meal, feces, or other animal-origin matter, because they view the production of these materials as either harming animals directly, or as being associated with the exploitation and consequent suffering of animals.

Some of these materials are by-products of animal husbandry, created during the process of cultivating animals for the production of meat, milk, skins, furs, entertainment, labour, or companionship; the sale of by-products decreases expenses and increases profit for those engaged in animal husbandry, and therefore helps support the animal husbandry industry, an outcome most vegans find unacceptable.

## Types

### *Forest Gardening*

Forest gardening is a fully plant-based organic food production system based on woodland ecosystems, incorporating fruit and nut trees, shrubs, herbs, vines and perennial vegetables. Making use of companion planting, these can be intermixed to grow in a succession of layers, to replicate a woodland habitat. Forest gardening can be viewed as a way to recreate the Garden of Eden. The three main products from a forest garden are fruit, nuts and green leafy vegetables.

Robert Hart adapted forest gardening for temperate zones during the early 1960s. Robert Hart began with a conventional smallholding at Wenlock Edge in Shropshire.

However, following his adoption of a raw vegan diet for health and personal reasons, Hart replaced his farm animals with plants. He created a model forest garden from a small orchard on his farm and intended naming his gardening method *ecological horticulture* or *ecocultivation*. Hart later dropped these terms once he became aware that *agroforestry* and *forest gardens* were already being used to describe similar systems in other parts of the world.

### *Vegan Permaculture*

Permaculture is an approach to designing human settlements and agricultural systems modelled on the relationships found in nature. It is based on three core values; "Earth Care, People Care and Fair Share." Animals are used as a common design element in permaculture. For example, chickens can be used as a method of weed control and for the production of eggs, meat and fertilizer. Bill Mollison, who coined the term permaculture, stated that "animals represent a valid method of storing inedible vegetation as food."

Vegan permaculture (also known as veganic permaculture, veganiculture or vegaculture) avoids the use of domesticated animals. It is essentially the same as permaculture except for the addition of a fourth core value; "Animal Care." Zalan Glen, a raw vegan, proposes that *vegaculture* should emerge out of permaculture in the same way veganism split from vegetarianism in the 1940s.

### *Veganic Gardening*

The veganic gardening method is a distinct system developed by Rosa Dalziell O'Brien, Kenneth Dalziel O'Brien and May E. Bruce. Although the term was originally coined by Geoffrey Rudd as a contraction of *vegetable organic* in order to "denote a clear distinction between conventional chemical based systems and organic ones based on animal manures". The O'Brien system's principal argument is that animal manures are harmful to soil health rather than that their use involves exploitation of and cruelty to animals.

The system employs very specific techniques including the addition of straw and other vegetable wastes to the soil in order to maintain soil fertility. Gardeners following the system use soil-covering mulches, and employ non-compacting surface cultivation techniques using any short-handled, wide-bladed, hand hoe. They kneel when surface cultivating, placing a board under their knees to spread out the pressure, and prevent soil compaction. Kenneth Dalziel O'Brien published a description of his system in *Veganic Gardening, the Alternative System for Healthier Crops*:

The veganic method of clearing heavily infested land is to take advantage of a plant's tendencies to move its roots nearer to the soil's surface when it is deprived of light. To make use of this principle, aided by a decaying process of the top growth of weeds, etc., it is necessary to subject such growth to heat and moisture in order to speed up the decay, and this is done by applying lime, then a heavy straw cover, and then the herbal compost activator...The following are required: Sufficient new straw to cover an area to be cleared to a depth of 3 to 4 inches.

Also part of the O'Brien method is minimal disturbance of the soil by tilling, use of compost, mulch, cover crops, and green manures, use of permanent raised beds and permanent hard-packed paths between them, laying out beds from north to south, putting plants in double rows or more so that not every row has a path on both sides.

### *Practices*

Soil fertility is maintained by the use of green manures and composted vegetable matter and minerals. Vegan gardeners may supplement this with human urine (which provides nitrogen) and 'humanure' produced from compost toilets.

Veganic gardeners may prepare soil for cultivation using the same method used by conventional and organic gardeners of breaking up the soil with hand tools and power tools and allowing the weeds to decompose. Shallow cultivation is also becoming popular among such gardeners: shallowly turning the soil's surface using one of numerous surface cultivating tools that are now available in the market place, including the popular Coleman surface-hoes developed by Eliot Coleman. Shallow tilling disturbs the soil less than deep turning and, combined with avoidance of soil compaction as described by O'Brien, helps the soil maintain ecological balance. This minimises the opportunities for soil diseases, plant diseases, and insect pests to become abundant.

## Environmental Vegetarianism

Environmental vegetarianism is the practice of vegetarianism or veganism based on the indications that animal production, particularly by intensive agriculture, is environmentally unsustainable. The primary environmental concerns with animal products are pollution and the use of resources such as fossil fuels, water, and land. Industrial monoculture is harvesting large quantities of a single food species, such as maize, or cattle. Monoculture is commonly practiced in industrial agriculture, which is more environmentally damaging than sustainable farming practices such as organic farming, permaculture, arable, pastoral, and rain-fed agriculture .

According to a 2006 Food and Agriculture Organisation report, industrialized agriculture contributes on a "massive scale" to climate change, air pollution, land degradation, energy use, deforestation, and biodiversity decline . The FAO report estimates that the livestock (including poultry) sector (which provides draft animal power, leather, wool, milk, eggs, fertilizer, pharmaceuticals, etc., in addition to meat) contributes about 18 percent of global GHG (greenhouse gas) emissions expressed as 100-year $CO_2$ equivalents.

This estimate was based on life cycle analysis, including feed production, land use changes, etc., and used GWP (global warming potential) of 23 for methane and 296 for nitrous oxide, to convert emissions of these gases to 100-year $CO_2$ equivalents. Some sources disagree with some of the figures used in arriving at the FAO estimate of 18 percent. For example, the FAO report estimates that 37 percent of global anthropogenic methane emissions are attributable to the livestock sector, and a US NASA summary indicates about 30 percent. Because of the GWP multiplier used, such a difference between estimates will have a large effect on an estimate of GHG $CO_2$ equivalents contributed by the livestock sector.

Livestock sources (including enteric fermentation and manure) account for about 3.1 percent of US anthropogenic GHG emissions expressed as $CO_2$ equivalents. This estimate is based on methodologies agreed to by the Conference of Parties of the UN FCCC. Data of a USDA study indicate that about 0.9 percent of energy use in the United States is accounted for by raising food-producing livestock and poultry. In this context, energy use includes energy from fossil, nuclear, hydroelectric, biomass, geothermal, technological solar, and wind sources. The estimated energy use in agricultural production includes embodied energy in purchased inputs. .

Another agricultural effect is on land degradation. Much of the world's crops is used to feed animals.. With 30 percent of the earth's land devoted to raising livestock, a major cutback is needed to keep up with growing population. A 2010 UN report explained that Western dietary preferences for meat would be unsustainable as the world population rose to the forecasted 9.1 billion by 2050. Demand for meat is expected to double by this date; meat consumption is steadily rising in countries such as China that once followed more sustainable, vegetable-based diets.

The environmental impacts of animal production vary with the method of production. A grazing-based production can limit soil erosion and also allow farmers to control pest problems with less pesticides through rotating crops with grass. In arid areas, however, it may catalyse a desertification process. The ability of soil to absorb water by infiltration is important for minimising runoff and soil erosion.

Researchers in Iowa reported that a soil under perennial pasture grasses grazed by livestock was able to absorb far more water than the same kind of soil under two annual crops: corn and soybeans. Corn and soybean crops commonly provide food for human consumption, biofuels, livestock feed, or some combination of these. The FAO initiative concluded that "the livestock sector emerges as one of the top two or three most significant contributors to the most serious environmental problems, at every scale from local to global."

### *Related Economic and Social Considerations*

Environmental vegetarianism can be compared with economic vegetarianism. An economic vegetarian is someone who practices vegetarianism either out of necessity or because of a conscious simple living strategy. Such a person may base this belief on a philosophical viewpoint, such as the belief that the consumption of meat is economically unsound or that vegetarianism will help improve public health and curb starvation. According to the Worldwatch Institute, "massive reductions in meat consumption in industrial nations will ease the health care burden while improving public health; declining livestock herds will take pressure off rangelands and grainlands, allowing the agricultural resource base to rejuvenate. As populations grow, lowering meat consumption worldwide will allow more efficient use of declining per capita land and water resources, while at the same time making grain more affordable to the world's chronically hungry."

Environmental vegetarians call for a reduction of first world consumption of meat, especially in the US. According to the United Nations Population Fund "Each U.S. citizen consumes an average of 260 lbs of meat per year, the world's highest rate. That is about 1.5 times the industrial world average, three times the East Asian average, and 40 times the average in Bangladesh." In addition, "the ecological footprint of an average person in a high-income country is about six times bigger than that of someone in a low-income country, and many more times bigger than in the least-developed countries."

The World Health Organisation calls malnutrition "the silent emergency", and says it is a factor in at least half of the 10.4 million child deaths which occur every year. Some argue that the adoption of a lacto-ovo vegetarian or entirely plant-based vegan diet is best, but may not be totally necessary, because even modest reductions in meat consumption in industrialized societies would substantially reduce the burden on our natural resources. For developed countries, a CAST report estimates an average of 2.6 pounds of grain fed per pound of beef carcass meat produced. For developing countries, the estimate is 0.3 pounds per pound. (Some very dissimilar figures are sometimes seen; the CAST report discusses common sources of error and discrepancies among such figures.) In 2007, US per capita beef consumption was 62.2 pounds per year, and US per capita meat (red meat plus fish plus poultry) consumption totalled 200.7 pounds, boneless trimmed weight basis.

A 2010 report from United Nations Environment Program's (UNEP) International Panel of Sustainable Resource Management, states that global shift towards a vegan diet is critical for mitigating global issues of hunger, fuel poverty and the worst impacts of climate change. The panel declared: "Impacts from agriculture are expected to increase substantially due to population growth and increasing consumption of animal products. Unlike fossil fuels, it is difficult to look for alternatives: people have to eat. A substantial reduction of impacts would only be possible with a substantial worldwide diet change, away from animal products."

A widely adopted vegetarian diet, in and of itself, may not be enough to make the US food system sustainable, unless greener agricultural practices, such as the adoption of renewable energy, are also implemented.

The support of alternative farming practices (e.g. well husbanded organic farming, permaculture, and rotational grazing) and the avoidance of certain plant commodities such as rice, also have a beneficial impact on environmental health and sustainable agriculture, though this would have little effect on animal welfare and rights. According to Cornell scientists, "the heavy dependence on fossil energy suggests that the US food system, whether meat-based or plant-based, is not sustainable." But they also mention that: "lactoovovegetarian diet is more sustainable than the average American meat-based diet."

Some environmental activists claim that adopting a vegetarian diet may be a way of focusing on personal actions and righteous gestures rather than systemic change. Dave Riley, an Australian environmentalist states that "being meatless and guiltless seems seductively simple while environmental destruction rages around us," noting that animals can contribute to the food chain. "For instance, yams, which keep poorly, are stored inside pigs, and today's rotting apples attracting fruit flies are tomorrow's bacon,".

Bill Mollison has inconsistently argued in his Permaculture Design Course that vegetarianism exacerbates soil erosion. This is because removing a plant from a field removes all the nutrients it obtained from the soil, while removing an animal leaves the field intact.

However, as feeding cattle requires many more plants to be removed from a field than direct harvesting for human consumption, this argument holds no ground. Furthermore, cattle are a known cause for soil erosion through trampling of the ground and overgrazing. However, on US farmland, much less soil erosion is associated with pastureland used for livestock grazing than with land used for production of crops. Robert Hart has also developed forest gardening, which has since been adopted as a common permaculture design element, as a sustainable plant-based food production system.

## Deep Ecology

Deep ecology is a contemporary ecological philosophy distinguished by its advocacy of the inherent worth of living beings regardless of their instrumental utility to human needs. Deep ecology argues that the natural world is a subtle balance of complex inter-relationships in which the existence of organisms are dependent on the existence of others within ecosystems. Human interference with or destruction of the natural world poses a threat therefore not only to humans but to all organisms constituting the natural order. Deep ecology's core principle is the belief that the living environment as a whole should be respected and regarded as having certain legal rights to live and flourish. It describes itself as "deep" because it regards itself as looking more deeply into the actual reality of humanity's relationship with the natural world arriving at philosophically more profound conclusions than that of the prevailing view of ecology as a branch of Darwinian biological science.

The movement does not subscribe to anthropocentric environmentalism (which is concerned with conservation of the environment only for exploitation by and for human purposes) since Deep ecology is grounded in a quite different set of philosophical assumptions. Deep ecology takes a more holistic view of the world human beings live in and seeks to apply to life the understanding that the separate parts of the ecosystem (including humans) function as a whole. This philosophy provides a foundation for the environmental, ecology and green movements and has fostered a new system of environmental ethics advocating wilderness preservation, human population control and simple living.

Deep ecology has been described by its critics as representing a politically and philosophically radical branch of environmentalism, profoundly misanthropic, regarding humanity as a pathological infestation on the earth.

Proponents of deep ecology believe that the world does not exist as a resource to be freely exploited by humans. The ethics of deep ecology hold that a whole system is superior to any of its parts. Proponents of deep ecology offer an eight-tier platform to elucidate their claims:

1. The well-being and flourishing of human and nonhuman life on Earth have value in themselves (synonyms: intrinsic value, inherent value). These values are independent of the usefulness of the nonhuman world for human purposes.
2. Richness and diversity of life forms contribute to the realisation of these values and are also values in themselves.
3. Humans have no right to reduce this richness and diversity except to satisfy vital human needs.
4. The flourishing of human life and cultures is compatible with a substantial decrease of the human population. The flourishing of nonhuman life requires such a decrease.
5. Present human interference with the nonhuman world is excessive, and the situation is rapidly worsening.
6. Policies must therefore be changed. These policies affect basic economic, technological, and ideological structures. The resulting state of affairs will be deeply different from the present.
7. The ideological change is mainly that of appreciating life quality (dwelling in situations of inherent value) rather than adhering to an increasingly higher standard of living. There will be a profound awareness of the difference between big and great.
8. Those who subscribe to the foregoing points have an obligation directly or indirectly to try to implement the necessary changes.

These principles can be refined down into three simple propositions:

1. Wilderness preservation;
2. Human population control;
3. Simple living (or treading lightly on the planet).

The phrase "deep ecology" was coined by the Norwegian philosopher Arne Næss in 1973. and he helped give it a theoretical foundation. Næss rejected the idea that beings can be ranked according to their relative value. For example, judgments on whether an animal has an eternal soul, whether it uses reason or whether it has consciousness (or indeed higher consciousness) have all been used to justify the ranking of the human animal as superior to other animals. Næss states that from an ecological point of view "the right of all forms [of life] to live is a universal right which cannot be quantified. No single species of living being has more of this particular right to live and unfold than any other species."

This metaphysical idea is elucidated in Warwick Fox's claim that humanity and all other beings are "aspects of a single unfolding reality". As such Deep Ecology would support the view of Aldo Leopold in his book, *A Sand County Almanac* that humans are "plain members of the biotic community". They also would support Leopold's "Land Ethic": "a thing is right when it tends to preserve the integrity, stability and beauty of the biotic community. It is wrong when it tends otherwise." Daniel Quinn in *Ishmael*, showed that an anthropocentric myth underlies our current view of the world, and a jellyfish would have an equivalent jellyfish centric view.

Deep ecology offers a philosophical basis for environmental advocacy which may, in turn, guide human activity against perceived self-destruction. Deep ecology and environmentalism hold that the science of ecology shows that ecosystems can absorb only limited change by humans or other dissonant influences. Further, both hold that the actions of modern civilization threaten global ecological well-being. Ecologists have described change and stability in ecological systems in various ways, including homeostasis, dynamic equilibrium, and "flux of nature". Regardless of which model is most accurate, environmentalists contend that massive human economic activity has pushed the biosphere far from its "natural" state through reduction of biodiversity, climate change, and other influences. As a consequence, civilization is causing mass extinction. Deep ecologists hope to influence social and political change through their philosophy. Næss has proposed, as Nicholas Goodrick-Clarke writes, "that the earth's human population should be reduced to about 100 million."

Næss and Fox do not claim to use logic or induction to derive the philosophy directly from scientific ecology but rather hold that scientific ecology directly implies the metaphysics of deep ecology, including its ideas about the self and further, that deep ecology finds scientific underpinnings in the fields of ecology and system dynamics.

In their 1985 book *Deep Ecology*, Bill Devall and George Sessions describe a series of sources of deep ecology. They include the science of ecology itself, and cite its major contribution as the rediscovery in a modern context that "everything is connected to everything else." They point out that some ecologists and natural historians, in addition to their scientific viewpoint, have developed a deep ecological consciousness—for some a political consciousness and at times a spiritual consciousness. This is a perspective beyond the strictly human viewpoint, beyond anthropocentrism. Among the scientists they mention specifically are Rachel Carson, Aldo Leopold, John Livingston, Paul R. Ehrlich and Barry Commoner, together with Frank Fraser Darling, Charles Sutherland Elton, Eugene Odum and Paul Sears.

A further scientific source for deep ecology adduced by Devall and Sessions is the "new physics." Which they describe as shattering Descartes's and Newton's vision of the universe as a machine explainable in terms of simple linear cause and effect. They propose that Nature is in a state of constant flux and reject the idea of observers as existing independent of their environment. They refer to Fritjof Capra's *The Tao of Physics* and *The Turning Point* for their characterisation of how the new physics leads to metaphysical and ecological views of interrelatedness, which, according to Capra, should make deep ecology a framework for future human societies. Devall and Sessions also credit the American poet and social critic Gary Snyder—with his devotion to Buddhism, Native American studies, the outdoors, and alternative social movements—as a major voice of wisdom in the evolution of their ideas.

The scientific version of the Gaia hypothesis was also an influence on the development of deep ecology. The central spiritual tenet of deep ecology is that the human species is a part of the Earth and not separate from it. Human life is dependent on the harmonious balance of interdependent relationships between organisms. In the words of Chief Seattle, "Man did not weave the web of life; he is merely a strand in it". A process of self-realisation or "re-earthing" is used for an individual to intuitively gain an ecocentric perspective. The notion is based on the idea that the more we *expand the self* to identify with "others" (people, animals, ecosystems), the more we realise ourselves. Transpersonal psychology has been used by Warwick Fox to support this idea. A number of spiritual and philosophical traditions including Native American, Buddhist and Jain are drawn upon in a continuing critique of the philosophical assumptions of the modern European mind which has enabled and led to what is seen as an increasingly unsustainable level of disregard to towards the rights and needs of the natural world and its ability to continue to support human life. In relation to the Judeo-Christian tradition, Næss offers the following criticism: "The arrogance of stewardship [as found in the Bible] consists in the idea of superiority which underlies the thought that we exist to watch over nature like a highly respected middleman between the Creator and Creation." This theme had been expounded in Lynn Townsend White, Jr.'s 1967 article "The Historical Roots of Our Ecological Crisis", in which however he also offered as an alternative Christian view of man's relation to nature that of Saint Francis of Assisi, who he says spoke for the equality of all creatures, in place of the idea of man's domination over creation. Næss' criticism is a criticism of the modern Protestant view of creation as property to be put into maximum productive use: a view used frequently in the past to exploit and dispossess native populations.

The original Christian teachings on property support the Franciscan/stewardship interpretation of the Bible. Against this view, Martin Luther condemned church ownership of lands because "they did not want to use that property in an economically productive fashion. At best they used it to produce prayers. Luther, and other Reformation leaders insisted that it should be used, not to relieve men from the necessity of working, but as a tool for making more goods. The attitude of the Reformation was practically, "not prayers, but production." And production, not for consumption, but for more production." This justification was offered to support secular takings of church endowments and properties.

Contemporary teaching of the Catholic Church expressly supports deep ecology themes: "Use of the mineral, vegetable, and animal resources of the universe cannot be divorced from respect for moral imperatives. Man's dominion over inanimate and other living beings granted by the Creator is not absolute; it is limited by concern for the quality of life or his neighbour, including generations to come; it requires a religious respect for the integrity of creation." This continues a traditional Catholic understanding of nature as a creation of God, and a tangible expression of God. Drawing upon the Buddhist tradition is the work of Joanna Macy. Macy, working as an anti-nuclear activist in the USA, found that one of the major impediments confronting the activists' cause was the presence of unresolved emotions of despair, grief, sorrow, anger and rage.

The denial of these emotions led to apathy and disempowerment. We may have intellectual understanding of our interconnectedness, but our culture, experiential deep ecologists like John Seed argue, robs us of emotional and visceral experience of that interconnectedness which we had as small children, but which has been socialised out of us by a highly anthropocentric alienating culture. Through "Despair and Empowerment Work" and more recently "The Work that Reconnects", Macy and others have been taking Experiential Deep Ecology into many countries including especially the USA, Europe (particularly Britain and Germany), Russia and Australia. The Dragon Dreaming approach developed by the Gaia Foundation of Australia for organisation and project planning is based upon experiential deep ecology.

### Spinoza

Arne Næss, who first wrote about the idea of deep ecology, from the early days of developing this outlook conceived Spinoza as a philosophical source. Others have followed Naess' inquiry, including Eccy de Jonge, in *Spinoza and Deep Ecology: Challenging Traditional Approaches to Environmentalism*, and Brenden MacDonald, in *Spinoza, Deep Ecology, and Human Diversity—Realisation of Eco-Literacies*

One of the topical centres of inquiry connecting Spinoza to Deep Ecology is "self-realisation."

### Savitri Devi

According to professor Nicholas Goodrick-Clarke, Deep ecology owes a major philosophical debt to Savitri Devi (1905–1982), a pioneer of the modern environmentalism and animal rights movements. Devi was also a passionate devotee of Adolf Hitler, fusing National Socialism with Hinduism and regarding Hitler as an Avatar of Vishnu who would destroy Judaeo-Christianity, Western culture and capitalism. Goodrick-Clarke writes:

> *The American movement of Deep Ecology betrays an uneasy resemblance to Savitri Devi's biocentric vision. Its precepts of community and cooperation are belied by romantic irrationalism and the assertion that all nature is equal. [Deep ecology and Devi both agree that] [o]nly a radical revaluation of human importance can avert the grim future of a teeming, polluted planet. Man must give up his privileged position as the lord of the earth and seek a new accommodation with nature, at once harmonious, modest, and subordinate. [...] Dave Foreman, the founder of the radical movement "Earth First!" conflates romanticism and brutality in a manner reminiscent of Savitri Devi...*

### *Movement*

In practice, deep ecologists support decentralization, and identifying politically with ecoregions; the breakdown of industrialism in its current form; and an end to authoritarianism. Deep ecology is not normally considered a distinct movement, but as part of the green movement. The deep ecological movement could be defined as those within the green movement who hold deep ecological views. Deep ecologists welcome the labels "Gaian" and "Green" (including the broader political implications of this term, e.g. commitment to peace). Deep ecology has had a broad general influence on the green movement by providing an independent ethical platform for Green parties, political ecologists and environmentalists. The philosophy of deep ecology helped differentiate the modern ecology movement by pointing out the anthropocentric bias of the term "environment", and rejecting the idea of humans as authoritarian guardians of the environment.

### *Interests in Nature*

Animal rights activists state that for something to require rights and protection intrinsically, it must have interests. Deep ecology is criticised for assuming that plants, for example, have their own interests as they are manifested by the plant's behaviour - self-preservation being considered an expression of a will to live, for instance. Deep ecologists claim to *identify* with non-human nature, and in doing so, criticise those who claim that non-human (or non-sentient) lifeforms' needs or interests are nonexistent or unknowable. The criticism is that the interests that a deep ecologist attributes to non-human organisms such as survival, reproduction, growth and prosperity are really human interests.

"The earth is endowed with 'wisdom', wilderness equates with 'freedom', and life forms are said to emit 'moral' qualities." However, the overarching criticism assumes that humans, in governing their own affairs, are somehow absent from this same assumption; i.e. how can governing humans truly presume to understand the interests of the rest of humanity. While the deep ecologist critic would answer that the logical application of language and social mores would provide this justification, i.e. voting patterns etc., the deep ecologist would note that these needs and interests are ultimately deducible solely from the application of logic to the observed behaviour of the life form, which is the same standard used by deep ecologists to perceive the standard of interests for the natural world.

Deep ecology is criticised for its claim to be *deeper* than alternative theories, which by implication are *shallow*. However despite repeated complaints about use of the term it still enjoys wide currency; *deep* evidently has an attractive resonance for many who seek to establish a new ethical framework for guiding human action with respect to the natural world. It may be presumptuous to assert that one's thinking is deeper than others'. When Arne Næss coined the term *deep ecology* he compared it favourably with *shallow environmentalism* which he criticized for its utilitarian and anthropocentric attitude to nature and for its materialist and consumer-oriented outlook. Against this is Arne Næss's own view that the "depth" of deep ecology resides in the persistence of its interrogative questioning, particularly in asking "Why?" when faced with initial answers.

### *Ecofeminist Response*

Both ecofeminism and deep ecology put forward a new conceptualization of the self. Some ecofeminists, such as Marti

Kheel, argue that self-realisation and identification with all nature places too much emphasis on the whole, at the expense of the independent being. Similarly, some ecofeminists place more emphasis on the problem of androcentrism rather than anthropocentrism. To others, like Karen J. Warren, the domination of women is tethered conceptually and historically to the domination of nature. Ecofeminism denies abstract individualism and embraces the interconnectedness of the living world; relationships, including our relationship with non-human nature, are not extrinsic to our identity and are essential in defining what it means to be human. Warren argues that hierarchical classifications in general, such as racism or speciesism, are all forms of discrimination and are no different than sexism. Thus, anthropocentrism is simply another form of discrimination as a result of our flawed value structure and should be abolished.

### *Misunderstanding Scientific Information*

Daniel Botkin has likened deep ecology to its antithesis, the wise use movement, when he says that they both "misunderstand scientific information and then arrive at conclusions based on their misunderstanding, which are in turn used as justification for their ideologies. Both begin with an ideology and are political and social in focus." Elsewhere though, he asserts that deep ecology must be taken seriously in the debate about the relationship between humans and nature because it challenges the fundamental assumptions of Western philosophy. Botkin has also criticized Næss's restatement and reliance upon the balance of nature idea and the perceived contradiction between his argument that all species are morally equal and his disparaging description of pioneering species.

### *"Shallow" View Superior*

Writer William Grey believes that developing a non-anthropocentric set of values is "a hopeless quest" He seeks an improved "shallow" view, writing, "What's wrong with shallow views is not their concern about the well-being of humans, but that they do not really consider enough in what that well-being consists. We need to develop an enriched, fortified anthropocentric notion of human interest to replace the dominant short-term, sectional and self-regarding conception."

### *Deep Ecology as Not "Deep" Enough*

Social ecologists such as Murray Bookchin claim that deep ecology fails to link environmental crises with authoritarianism and hierarchy. Social ecologists believe that environmental problems are firmly rooted in the manner of human social interaction, and protest that an ecologically sustainable society could still be socially exploitative. Deep ecologists reject the argument that ecological behaviour is rooted in the social paradigm (according to their view, that is an anthropocentric fallacy), and they maintain that the converse of the social ecologists' objection is also true in that it is equally possible for a socially egalitarian society to continue to exploit the Earth.

Parallels have been drawn between deep ecology and other philosophies, in particular those of the animal rights movement and Earth First! Peter Singer's 1975 book *Animal Liberation* critiqued anthropocentrism and put the case for animals to be given moral consideration. This can be seen as a part of a process of expanding the prevailing system of ethics to wider groupings. However, Singer has disagreed with deep ecology's belief in the intrinsic value of nature separate from questions of suffering, taking a more utilitarian stance. The feminist and civil rights movements also brought about expansion of the ethical system for their particular domains. Likewise deep ecology brought the whole of nature under moral consideration. The links with animal rights are perhaps the strongest, as "proponents of such ideas argue that 'All life has intrinsic value'". Many in the radical environmental direct-action movement Earth First! claim to follow deep ecology, as indicated by one of their slogans *No compromise in defence of mother earth*. In particular, David Foreman, the co-founder of the movement, has also been a strong advocate for deep ecology, and engaged in a public debate with Murray Bookchin on the subject. Judi Bari was another prominent Earth Firster who espoused deep ecology. Many Earth First! actions have a distinct deep ecological theme; often these actions will be to save an area of old growth forest, the habitat of a snail or an owl, even individual trees. Actions are often symbolic or have other political aims. At one point Arne Næss also engaged in environmental direct action, though not under the Earth First! banner, when he tied himself to a Norwegian fjord in a successful protest against the building of a dam. There are also anarchist currents in the movement, especially in the United Kingdom. For example Robert Hart, pioneer of forest gardening in temperate climates, wrote the essay "Can Life Survive?" in *Deep Ecology & Anarchism*.

Robert Greenway and Theodore Roszak have employed the deep ecology platform as a means to argue for ecopsychology. Although ecopsychology is a highly differentiated umbrella that encompasses many practices and perspectives, its ethos is generally consistent with deep ecology. As this now almost forty-year old "field" expands and continues to be reinterpreted by a variety of practitioners, social and natural scientists, and humanists, "ecopsychology" may change to include these novel perspectives.

## Plants for a Future

Plants For A Future (PFAF) is an online not for profit resource for those interested in edible and useful plants of temperate regions. The project currently has a site in the South West of England where many of the plants are being grown on a trial basis, and maintains a small mail order catalogue. The organisation's emphasis is on perennial plants.

PFAF is a registered educational charity with strong ethical principals and the following objectives:

The Charity's objectives are to advance the education of the public by the promotion of all aspects of ecologically sustainable vegan-organic horticulture and agriculture with an emphasis on tree, shrub and other perennial species; and the undertaking of research into such horticulture and agriculture, and dissemination of the results of such research.

## Synergistic Gardening

Synergistic gardening is a system of organic gardening, developed by Emilia Hazelip, and based on the work of Masanobu Fukuoka. The approach is a type of permaculture. After establishing the garden, there is no further digging, ploughing or tilling, and no use of external inputs such as manures and other fertilizers, or pesticides. Soil health is maintained by the selection of plants, mulching, and recycling of plant residues.

## Forest Gardening

Forest gardening is a low-maintenance sustainable plant-based food production and agroforestry system based on woodland ecosystems, incorporating fruit and nut trees, shrubs, herbs, vines and perennial vegetables which have yields directly useful to humans. Making use of companion planting, these can be intermixed to grow in a succession of layers, to replicate a woodland habitat.

Forest gardens are probably the world's oldest form of land use and most resilient agroecosystem. They originated in prehistoric times along jungle-clad river banks and in the wet foothills of monsoon regions. In the gradual process of families improving their immediate environment, useful tree and vine species were identified, protected and improved whilst undesirable species were eliminated. Eventually superior foreign species were selected and incorporated into the gardens.

Forest gardens are still common in the tropics and known by various names such as: *home gardens* in Kerala in South India, Nepal, Zambia, Zimbabwe and Tanzania; *Kandyan forest gardens* in Sri Lanka; *huertos familiares*, the "family orchards" of Mexico; and *pekarangan*, the gardens of "complete design", in Java. These are also called agroforests and, where the wood components are short statured, the term shrub garden is employed. Forest gardens have been shown to be a significant source of income and food security for local populations.

### *In Tropical Climates*

Forest gardens, or home gardens, are common in the tropics, using inter-cropping to cultivate trees, crops, and livestock on the same land. In Kerala in south India as well as in northeastern India, the home garden is the most common form of land use and is also found in Indonesia.

One example combines coconut, black pepper, cocoa and pineapple. These gardens exemplify polyculture, and conserve much crop genetic diversity and heirloom plants that are not found in monocultures. Forest gardens are loosely connected to the religious concept of the Garden of Eden.

The BBC's *Unnatural Histories* claimed that the Amazon rainforest, rather than being a pristine wilderness, has been shaped by man for at least 11,000 years through practices such as forest gardening and terra preta. Since the 1970s, numerous geoglyphs have also been discovered on deforested land in the Amazon rainforest, furthering the evidence about Pre-Columbian civilizations.

On the Yucatán Peninsula, much of the Maya food supply was grown in "orchard-gardens", known as *pet kot*. The system takes its name from the low wall of stones (*pet* meaning circular and *kot* wall of loose stones) that characteristically surrounds the gardens.

In many African countries, for example Zambia, Zimbabwe, Tanzania, gardens are widespread in rural, periurban and urban areas and they play an essential role in establishing food security. Most well known are the Chaga or Chagga gardens on the slopes of Mt. Kilimanjaro in Tanzania. These are an excellent example of an agroforestry system. In many countries, women are the main actors in home gardening and food is mainly produced for subsistence. In North-Africa, oasis layered gardening with palm trees, fruit trees and vegetables is a traditional type of forest garden.

Home gardens in Kerala are connected to 'Kaavu'. Some of the endangered species of wild plants, trees animals and plants can be seen in Home Gardens of Kerala. The main aim of preserving a Home Garden is to preserve ritualistic beliefs and cultural identity of joint families. Gradually, due to the origin of nuclear families and lifestyle related to the same, the Home Gardens underwent transformation to playgrounds. But still in Kerala, some families show ample importance to protect their Home Gardens. As pointed out, the preservation of home gardens in Kerala is strictly connected to family purposes. At the same time, home gardens are helpful to raise the water table of nearby ponds and wells. Unknowingly, some families in Kerala which preserve their home gardens preserve nature and ecological balance.

Kannur is an incredible district in Kerala which shows utmost interest in preserving forest gardens. Besides, the forest gardens in Kannur are preserved with due care because of the traditional way of worshiping nature as the embodiment of God. For instance, the 'Kaavus' or the local temples in Kannur possess the glory of preserved forests without human encroachment. Some of the most endangered species of wild life can be seen in Kaavus. For instance, the 'Parappool Kaavu' (Mele Kaavu and Thazhe Kaavu) and 'Kayyath Naagam' are important local temples in Taliparamba (Kannur District) with private forests/ forest gardens. The 'Parappool Kaavu' is situated at Parappool, 4-5 kilometres away from Taliparamba.

The rich vegetation within the area of the Parappool Kaavu is still to be explored. On the other side, the Kayyath Naagam is closely connected to the Indian way of worshiping snakes. The vegetation preserved within the area includes 30 acres of land. The surrounding areas of Kayyam are full of paddy fields. The rich vegetation preserved within the area of Kayyath Naagam acts the role of a lung which purifies and recharges the air, water and the atmosphere. Normally, the believers are allowed to visit the temple not the preserved area. But, the believers are allowed to visit the preserved area on 'Aayilyam Naal' or a particular day in every year. Besides, the Kayyath Naagam authorities show keen interest to preserve the vegetation by not allowing the poachers to enter the preserved area.

In Nepal, the Ghar Bagaincha, literally Home Garden, refers to the traditional land use system around a homestead, where several species of plants are grown and maintained by household members and their products are primarily intended for the family consumption (Shrestha et al., 2002). The term "home garden" is often considered synonymous to the kitchen garden. However, they differ in terms of function, size, diversity, composition and features (Sthapit et al., 2006). In Nepal, 72% of households have home gardens of an area 2-11% of the total land holdings (Gautam et al., 2004).

Because of their small size, the government has never identified home gardens as an important unit of food production and it thereby remains neglected from research and development. However, at the household level the system is very important as it is the an important source of quality food and nutrition for the rural poor and, therefore, are important contributors to the household food security and livelihoods of farming communities in Nepal.

They are typically cultivated with a mixture of annual and perennial plants that can be harvested on a daily or seasonal basis. Biodiversity that has an immediate value is maintained in home gardens as women and children have easy access to preferred food, and for this reason alone we should promote home gardens as a key element for a healthy way of life. Home gardens, with their intensive and multiple uses, provide a safety net for households when food is scarce.

These gardens are not only important sources of food, fodder, fuel, medicines, spices, herbs, flowers, construction materials and income in many countries, they are also important for the in situ conservation of a wide range of unique genetic resources for food and agriculture (Subedi et al., 2004). Many uncultivated, as well as neglected and underutilised species could make an important contribution to the dietary diversity of local communities (Gautam et al., 2004).

In addition to supplementing diet in times of difficulty, home gardens promote whole-family and whole-community involvement in the process of providing food. Children, the elderly, and those caring for them can participate in this infield agriculture, incorporating it with other household tasks and scheduling. This tradition has existed in many cultures around the world for thousands of years.

### *In Temperate Climates*

Robert Hart adapted forest gardening for temperate zones during the early 1960s. Hart began farming at Wenlock Edge in Shropshire with the intention of providing a healthy and therapeutic environment for himself and his brother Lacon. Starting as relatively conventional smallholders, Hart soon discovered that maintaining large annual vegetable beds, rearing livestock and taking care of an orchard were tasks beyond their strength. However, a small bed of perennial vegetables and herbs he planted was looking after itself with little intervention. Following Hart's adoption of a raw vegan diet for health and personal reasons, he replaced his farm animals with plants. The three main products from a forest garden are fruit, nuts and green leafy vegetables.

He created a model forest garden from a 0.12 acre (500 $m^2$) orchard on his farm and intended naming his gardening method *ecological horticulture* or *ecocultivation.*

Hart later dropped these terms once he became aware that *agroforestry* and *forest gardens* were already being used to describe similar systems in other parts of the world. He was inspired by the forest farming methods of Toyohiko Kagawa and James Sholto Douglas, and the productivity of the Keralan home gardens as Hart explains:

From the agroforestry point of view, perhaps the world's most advanced country is the Indian state of Kerala, which boasts no fewer than three and a half million forest gardens...As an example of the extraordinary intensivity of cultivation of some forest gardens, one plot of only 0.12 hectares (0.30 acres) was found by a study group to have twenty-three young coconut palms, twelve cloves, fifty-six bananas, and forty-nine pineapples, with thirty pepper vines trained up its trees. In addition, the small holder grew fodder for his house-cow.

4

# Techniques of Organic Gardening and Farming

## Multiple Cropping

In agriculture, multiple cropping is the practice of growing two or more crops in the same space during a single growing season. It is a form of polyculture. It can take the form of double-cropping, in which a second crop is planted after the first has been harvested, or relay cropping, in which the second crop is started amidst the first crop before it has been harvested. A related practice, companion planting, is sometimes used in gardening and intensive cultivation of vegetables and fruits. One example of multi-cropping is tomatoes + onions + marigold; the marigolds repel some tomato pests.

Multiple cropping is found in many agricultural traditions. In the Garhwal Himalaya of India, a practice called *baranaja* involves sowing 12 or more crops on the same plot, including various types of beans, grams, and millets, and harvesting them at different times. In the cultivation of rice, multiple cropping requires effective irrigation, especially in areas with a dry season. Rain that falls during the wet season permits the cultivation of rice during that period, but during the other half of the year, water cannot be channeled into the rice fields without an irrigation system.

The Green Revolution in Asia led to the development of high-yield varieties of rice, which required a substantially shorter growing season of 100 days, as opposed to traditional varieties, which needed 150 to 180 days. Due to this, multiple cropping became more prevalent in Asian countries. One kind of multiple cropping is intercropping, where an additional crop is planted in the spaces available between the main crop.

## Underutilised Crop

Neglected and underutilised crops are domesticated plant species that have been used for centuries or even millennia for their food, fibre, fodder, oil or medicinal properties, but have been reduced in importance over time owing to particular supply and use constraints. These can inter alia include poor shelf life, un-recognised nutritional value, poor consumer awareness and reputational problems ("poor people's food"). As the demand for plant and crop attributes changes (re-appraisal or discovery of nutritional traits, culinary value, adaptation to climate change, etc.), neglected crops can overcome the constraints to the wider production and use.

As a matter of fact, many formerly neglected crops are now globally significant crops (oilpalm, soybean, kiwi fruit). Although the options for scaling-up neglected crops for large-scale agriculture appear to be increasingly exhausted, many species have the potential to contribute to food security, nutrition, dietary and culinary diversification, health and income generation. They also provide environmental services. It is impossible to define what would constitute "proper" or "correct" levels of utilisation, however, it is evident that many neglected species are under-utilised relative to their nutritional value and productivity.

Just three crops - maize, wheat and rice - account for about 40% of the world's consumption of calories and protein. 95% of the world's food needs are provided for by just 30 species of plants. In stark contrast, at least 12,650 species names have been compiled as edible (Kunkel 1984). Neglected and underutilised plants are those that could be - and, in many cases, historically have been - used for food and other uses on a larger scale. Such crop species have also been described as "minor", "orphan", "promising" and "little-used". They continue to play an important role in the subsistence and economy of poor people throughout the developing world, particularly in the agrobiodiversity-rich tropics. Despite their potential for dietary diversification and the provision of micro-nutrients such as vitamins and minerals, they continue to attract little research and development attention. Alongside their commercial potential, many of the underutilised crops also provide important environmental services, as they are adapted to marginal soil and climate conditions.

It is difficult to precisely define which attributes makes a crop "underutilised", but often they display the following features:

- Linkage with the cultural heritage of their places of origin
- Local and traditional crops whose distribution, biology, cultivation and uses are poorly documented
- Adaptation to specific agro-ecological niches and marginal land
- Weak or no formal seed supply systems
- Traditional uses in localized areas
- Produced in traditional production systems with little or no external inputs
- Receive little attention from research, extension services, policy and decision makers, donors, technology providers and consumers
- May be highly nutritious and/or have medicinal properties or other multiple uses.

## No-dig Gardening

No-dig gardening is a non-cultivation method used by some organic gardeners. The origins of no-dig gardening are unclear, and may be based on pre-industrial or nineteenth-century farming techniques. Two pioneers of the method in the twentieth century included F. C. King, Head Gardener at Levens Hall, South Westmorland, in the Lake District of England, who wrote the book "Is Digging Necessary?" in 1946 and a gardener from Middlecliffe in the UK, A. Guest, who in 1948 published the book "Gardening Without Digging". The work of these gardeners was supported by the Good Gardeners Association in the UK. Masanobu Fukuoka started his pioneering research work in this domain in 1938, and began publishing in the 1970s his Fukuokan philosophy of "Do Nothing Farming", which is now acknowledged by some as the tap root of the Permaculture movement. No-dig gardening was also promoted by Australian Esther Deans in the 1970s, and American gardener Ruth Stout advocated a "permanent" garden mulching technique in *Gardening Without Work* and no-dig methods in the 1950s and 1960s.

This technique recognises that micro- and macro-biotic organisms constitute a "food web" community in the soil, necessary for the healthy cycling of nutrients and prevention of problematic organisms and diseases. The plants transfer a portion of the carbon energy they produce to the soil, and microbes that benefit from this energy in turn convert available organic substances in the soil to the mineral elements the plants need to thrive.

Historically the reasons for tilling the soil are to remove weeds, loosen and aerate the soil, and incorporate organic matter such as compost or manure into lower soil layers. In areas with thin soil and high erosion there is a strong case against digging, which argues that in the long term it can be detrimental to the food web in the fragile topsoil. While digging is an effective way of removing perennial weed roots, it also often causes seeds that can remain dormant for many decades to come to the surface and germinate. Digging can also damage soil structure, causing compaction, and unbalance symbiotic and mutualist interactions among soil life. Digging tends to displace nutrients, shifting surface organic material deeper, where there is less oxygen to support the decomposition to plant-available nutrients, which then need to be otherwise replenished. Digging is practiced traditionally in countries with old, deep, rich soils such as Western Europe, however traditionally there, this is followed by periodic resting of the soil, usually with an undisturbed cover crop.

No-dig methods allow nature to carry out cultivation operations. Organic matter such as well rotted manure, compost, leaf mold, spent mushroom compost, old straw, etc., is added directly to the soil surface as a mulch at least 5-15 centimetres (2–6 in) deep, which is then incorporated by the actions of worms, insects and microbes. Worms and other soil life also assist in building up the soil's structure, their tunnels providing aeration and drainage, and their excretions bind together soil crumbs. This natural biosphere maintains healthy conditions in the upper soil horizons where annual plant roots thrive. No-dig systems are said to be freer of pests and disease, possibly due to a more balanced soil population being allowed to build up in this undisturbed environment, and by encouraging the buildup of beneficial rather than harmful soil fungi. Moisture is also retained more efficiently under mulch than on the surface of bare earth, allowing slower percolation and less leaching of nutrients.

Another no-dig method is sheet mulching wherein a garden area is covered with wetted paper or cardboard, compost and topped off with landscape mulch.

A no-dig system is easier than digging. It is a long term process, and is reliant upon having plentiful organic matter to provide mulch material. It is also helpful to remove any perennial weed roots from the area beforehand, although their hold can be weakened by applying a light-excluding surface layer such as large sheets of cardboard or several thicknesses of spread out newspaper before adding the compost mulch. The newspaper or cardboard should be thoroughly wet as well. A popular book, *Lasagna Gardening* by Patricia Lanza (Rodale Press, Inc.) provides excellent instructions for the novice user.

Esther Dean, an Australian gardener and author, pioneered no dig gardening. Dean wrote the books *No-Dig Gardening* and *Leaves of Life*. She was still actively gardening at age 86 and promoted gardens for those with special needs. She inspired many famous gardeners including Bill Mollison of the Permaculture Movement although she said “it is not quite how I would do it”, implying she did not allow nature to take over, but retained formality with strict garden edges and more annuals.

Gardens fashioned on Esther Dean’s no dig gardening principles include Randwick Community Organic Garden (RCOG), Sydney, New South Wales, Australia.

## No-till Farming

No-till farming (also called zero tillage or direct planting or pasture cropping) is a way of growing crops from year to year without disturbing the soil through tillage. No-till is an agricultural technique which increases the amount of water and organic matter (nutrients) in the soil and decreases erosion. It increases the amount and variety of life in and on the soil but may require herbicide usage.

Tilling is used to remove weeds, shape the soil into rows for crop plants and furrows for irrigation. This leads to unfavourable effects, like soil compaction; loss of organic matter; degradation of soil aggregates; death or disruption of soil microbes and other organisms including mycorrhiza, arthropods, and earthworms; and soil erosion where topsoil is blown or washed away.

No-till farming avoids these effects by excluding the use of tillage. With this way of farming, crop residues or other organic amenities are retained on the soil surface and sowing/fertilizing is done with minimal soil disturbance. Continuous no-till needs to be managed very differently in order to keep or increase yield on the field. Residue, weeds, equipment, crop rotations, water, disease, pests, and fertilizer management are just some of the many details of farming that change when switching to no-till.

The idea of modern no-till started in the 1940s with Edward Faulkner, but it wasn't until the development of several chemicals after WWII that various researchers and farmers started to try out the idea. The first adopters of no-till include Klingman (North Carolina), Edward Faulkner, L.A. Porter (New Zealand), Harry and Lawrence Young (Herndon, Kentucky), the Instituto de Pesquisas Agropecuarias Meridional (1971 in Brazil) with Herbert Bartz.

### *Profit, Economics, Yield*

Studies have found that no-till farming can be more profitable if performed correctly. Less tillage of the soil reduces labour, fuel, irrigation and machinery costs. No-till can increase yield because of higher water infiltration and storage capacity, and less erosion. Another benefit of no-till is that because of the higher water content, instead of leaving a field fallow it can make economic sense to plant another crop instead.

As sustainable agriculture becomes more popular, monetary grants and awards are becoming readily available to farmers who practice conservation tillage. Some large energy corporations which are among the greatest generators of fossil-fuel-related pollution are willing to purchase carbon credits to encourage farmers to engage in conservation tillage. The farmers' land essentially becomes a carbon sink for the power generators' emissions. This helps the farmer in several ways, and it helps the energy companies meet demands for reduction of pollution. No-till farming can increase organic (carbon based) matter in the soil as a form of carbon sequestration.

### *Carbon (Air and Soil) and Other Greenhouse Gases*

No-till has carbon sequestration potential through storage of soil organic matter in the soil of crop fields. Tilled by machinery, the soil layers invert, air mixes in, and soil microbial activity dramatically increases over baseline levels. The result is that soil organic matter is broken down much more rapidly, and carbon is lost from the soil into the atmosphere. This, in addition to the emissions from the farm equipment itself, increases carbon dioxide levels in the atmosphere.

Cropland soils are ideal for use as a carbon sink, since they have been depleted of carbon in most areas. It is estimated that 78 billion metric tonnes of carbon that was trapped in the soil has been released because of tillage. Conventional farming practices that rely on tillage have removed carbon from the soil ecosystem by removing crop residues such as left over corn stalks, and through the addition of chemical fertilizers which have the above mentioned effects on soil microbes.

By eliminating tillage, crop residues decompose where they lie, and by growing winter cover crops, carbon loss can be slowed and eventually reversed. In addition to keeping carbon in the soil, no-till farming reduces nitrous oxide ($N_2O$) emissions by 40-70%, depending on rotation. Nitrous oxide is a potent greenhouse gas that stays in the atmosphere for 120 years.

### *Soil and Water*

No-till improves soil quality (soil function), carbon, organic matter, aggregates, protecting the soil from erosion, evaporation of water, and structural breakdown. A reduction in tillage passes helps prevent the compaction of soil. Recently, researchers at the Agricultural Research Service found that no-till farming makes soil much more stable than plowed soil. Their conclusions draw from over 19 years of collaborated tillage studies. No-till stores more carbon in the soil and carbon in the form of organic matter is a key factor in holding soil particles together. The first inch of no-till soil is two to seven times less vulnerable than that of plowed soil. The practice of no-till farming is especially beneficial to Great Plains farmers because of its resistance to erosion. Crop residues left intact help both natural precipitation and irrigation water infiltrate the soil where it can be used. The crop residue left on the soil surface also limits evaporation, conserving water for plant growth. Soil compaction and no tillage-pan, soil absorbs more water and plants are able to grow their roots deeper into the soil and suck up more water. Tilling a field reduces the amount of water, via evaporation, around 1/3 to 3/4 inches (0.85 to 1.9 cm) per pass. By no-tilling, this water stays in the soil, available to the plants.

### *Soil Biota, Wildlife, etc.*

In no-till farming the soil is left intact and crop residue is left on the field. Therefore, soil layers, and in turn soil biota, are conserved in their natural state. No-tilled fields often have more beneficial insects and annelids, a higher microbial content, and a greater amount of soil organic material. Since there is no ploughing there is less airborne dust.

No-till increases the amount and variety of wildlife. This is the result of improved cover, reduced traffic and the reduced chance of destroying ground nesting birds and animals (plowing destroys all of them).

### *Albedo*

Tillage lowers the albedo of croplands. The potential for global cooling as a result of increased Albedo in no till croplands is similar in magnitude to the biogeochemical (carbon sequestration) potential.

### *Historical Artifacts*

Tilling regularly damages ancient structures under the soil such as long barrows. In the UK, half of the long barrows in Gloucestershire and almost all the burial mounds in Essex have been damaged. According to English Heritage modern tillage techniques have done as much damage in the last six decades as traditional tilling did in previous 6 centuries. By using no-till methods these structures can be preserved and can be properly investigated instead of being destroyed. Prior to no-till farming's rise in popularity, the annual tilling of the soil often exposed arrowheads and other artifacts. Other artifacts include bullets, medals, and buttons, coins and other metal items from destroyed houses and barns.

### *Cost*

No-till requires specialised seeding equipment designed to plant seeds into undisturbed crop residues and soil. Purchasing new equipment is expensive and while the cost could be offset by selling off plows, etc. doing so is not usually done until the farmer decides to switch completely over (after trying it out for a few years). This results in more money being invested into equipment in the short term (until old equipment is sold off).

### *Drainage*

If a soil has poor drainage, it may need drainage tiles or other devices in order to help with the removal of excess water under no-till. Farmers should remember that water infiltration will improve after several years of a field being in no-till, so they may want to wait until 5–8 years have passed to see if the problems persists before deciding to invest in such an expensive system.

### *Gullies*

Gullies can be a problem in the long-term. While much less soil is displaced by using no-till, any drainage gulleys that do form will get deeper each year since they aren't being smoothed out by plowing. This may necessitate either sod drainways, waterways, permanent drainways, cover crops, etc.

### *Increased Chemical Use*

One of the purposes of tilling is to remove weeds. No-till farming does change weed composition drastically. Faster growing weeds may no longer be a problem in the face of increased competition, but shrubs and trees may begin to grow eventually. Some farmers attack this problem with a "burn-down" herbicide such as glyphosate in lieu of tillage for seedbed preparation and because of this, no-till is often associated with increased chemical use in comparison to traditional tillage based methods of crop production. However, there are many agroecological alternatives to increased chemical use, such as winter cover crops and the mulch cover they provide, solarization or flaming. No-till requires some different skills in order to do it successfully. As with any production system, if no-till isn't done correctly, yields can drop. A combination of technique, equipment, pesticides, crop rotation, fertilization, and irrigation have to be used for local conditions.

### *Cover Crops*

Cover crops are used occasionally in no-till to help control weeds and increase nutrients in the soil (by using legumes) or by using plants with long roots to pull mobile nutrients back up to the surface from lower layers of the soil. Farmers experimenting with organic no-till use cover crops instead of tillage for controlling weeds, and are developing various methods to kill the cover crops (rollers, crimper, choppers, etc.) so that the newly planted crops can get enough light, water, nutrients, etc.

### *Disease, Pathogens, Insects and the use of Crop Rotations*

With no-till, residue from the previous years crops lie on the surface of the field, cooling it and increasing the moisture. This can cause increased or decreased or variations of diseases that occur, but not necessarily at a higher or lower rate than conventional tillage. In order to help eliminate weed, pest and disease problems, Crop rotations are used. By rotating the crops on a multi-year cycle, pests and diseases will decrease since the pests will no longer have a food supply to support their numbers.

### *Organic no-till Technique: The Cardboard Method*

Some farmers who prefer to pursue a chemical-free management practice often rely on the use of normal, non-dyed corrugated cardboard for use on seed-beds and vegetable areas. Used correctly, cardboard placed on a specific area can A) keep important fungal hyphae and microorganisms in the soil intact B) prevent recurring weeds from popping up C) increase residual nitrogen and plant nutrients by top-composting plant residues and D) create valuable topsoil that is well suited for next years seeds or transplants. The plant residues (left over plant matter originating from cover crops, grass clippings, original plant life etc.) will rot while underneath the cardboard so long as it remains sufficiently moist. This rotting attracts worms and other beneficial microorganisms to the site of decomposition, and over a series of a few seasons (usually Spring—>Fall or Fall—>Spring) and up to a few years, will create a layer of rich topsoil. Plants can then be direct seeded into the soil come spring, or holes can be cut into the cardboard to allow for transplantation. Using this method in conjunction with other sustainable practices such as composting/vermicompost, cover crops and rotations are often considered beneficial to both land and those who take from it.

### *Water Issues*

No-till dramatically reduces the amount of erosion in a field. While much less soil is displaced, any gullies that do form will get deeper each year instead of being smoothed out by regular plowing. This may necessitate either sod drainways, waterways, permanent drainways, cover crops, etc.

A problem that occurs in some fields is water saturation in soils. Switching to no-till will help the drainage issue because of the qualities of soil under continuous no-till includes a higher water infiltration rate. So, you might want to see how a field with saturated soil behaves under no-till before deciding whether or not to purchase tilling for the area.

### *Equipment*

It is very important to have planting equipment that can properly penetrate through the residue, into the soil and prepare a good seedbed. Switching to no-till reduces the maximum amount of power needed from farm tractors, which means that a farmer can farm under no-till with a smaller tractor than if he/she was tilling. Using a smaller, lighter tractor has the added benefit of reducing compaction.

### *Soil Temperature*

Another problem that growers face is that in the spring the soil will take longer to warm and dry, which may stall planting to a less ideal future date. One reason why the soil is slower to warm is that the field absorbs less solar energy as the residue covering the soil is a much lighter colour than the black soil which would be exposed in conventional tillage. This can be managed by using row cleaners on a planter. Since the soil can be cooler, harvest can occur a few days later than a conventionally tilled field. Note: A cooler soil is also a benefit because water doesn't evaporate as fast. On some crops, like continuous no-till corn, the thickness of the residue on the surface of the field can become a problem without proper preparation and/or equipment.

One of the most common yield reducers is because of fertilizer not being applied in slightly higher quantities during transition period, the time it takes the soil to rebuild its aggregates and organic matter. Yields will decrease temporarily because of nitrogen being immobilized in the crop residue, which can take a few months to several years to decompose, depending on the crop's C to N ratio and the local environment. The nutrients in the organic matter will be eventually released back into the soil, so this is only a concern during the transition time frame (4–5 years for Kansas, USA). This can be fixed by adding extra fertilizer during this time period.

Although no-till farming often causes a slight increase in soil bulk density, periodic tilling is not needed to "fluff" the soil back up. No-till farming mimics the natural conditions under which most soils formed more closely than any other method of farming, in that the soil is left undisturbed except to place seeds in a position to germinate.

No-till farming is not equivalent to conservation tillage or strip tillage. Conservation tillage is a group of practices that reduce the amount of tillage needed. No-till and strip tillage are both forms of conservation tillage. Remember, no-till is the practice of never tilling a field. Tilling every other year is called rotational tillage.

Research by both institutions and farmers continues into developing organic no-till farming methods that utilise the rolling/crimping of cover crops and diverse crop rotations to suppress weeds, insects, and diseases. Current organic farming methods often rely on tillage to control these pests at the expense of soil quality. The marriage of no-till and organic has the potential to produce both the healthiest food and the healthiest soil at the same time.

## Pollination Management

Pollination Management is the label for horticultural practices that accomplish or enhance pollination of a crop, to improve yield or quality, by understanding of the particular crop's pollination needs, and by knowledgeable management of pollenizers, pollinators, and pollination conditions.

### *Pollinator Decline*

With the decline of both wild and domestic pollinator populations, pollination management is becoming an increasingly important part of horticulture. Factors that cause the loss of pollinators include pesticide misuse, unprofitability of beekeeping for honey, rapid transfer of pests and diseases to new areas of the globe, urban/suburban development, changing crop patterns, clearcut logging (particularly when mixed forests are replaced by monoculture pine), clearing of hedgerows and other wild areas, loss of nectar corridors for migratory pollinators, and human paranoia of stinging insects (killer bee hype).

In 1989, following Hurricane Hugo, massive aerial applications for mosquitoes were done in South Carolina. The following year, watermelon growers who did not place beehives in the fields, observed the fruit begin to develop, then abort, or develop into small deformed fruit. There were entire fields that never yielded a single usable melon. Some growers went out of business; others began to seriously manage pollination. Since beekeepers were also heavily damaged by the mosquito spraying, the supply of bees for pollination was critically short for several years.

The term pollinator decline refers to the reduction in abundance of pollinators in many ecosystems worldwide during the end of the twentieth century. Pollinators participate in sexual reproduction of many plants, by ensuring cross-pollination, essential for some species, or a major factor in ensuring genetic diversity for others. Since plants are the primary food source for animals, the reduction of one of the primary pollination agents, or even their possible disappearance, has raised concern, and the conservation of pollinators has become part of biodiversity conservation efforts.

As plantings have grown larger, the need for concentrated pollinators at bloom time has grown. At the same time populations of many pollinators have been declining, and this decline has become a major environmental issue today. Pollination management seeks to protect, enhance, and augment agricultural pollination. For example, feral honey bee populations in the US have dropped about 90% in the past 50 years, except for the Southwest where they have been replaced by Africanized bees. At the same time managed honey bee colonies have dropped by about two thirds.

In North America, during the winter and spring of 2006 – 2007, was a considerable decline in commercially managed honeybee colonies, with losses of about one third of honeybees population. This event is named colony collapse disorder, and has appeared in 35 states of USA affecting honeybees, reporting losses in honeybees colonies as high as 80 to 100 percent for some beekeepers. Monoculture needs very high populations of bees at bloom, but can make the area quite barren, or even toxic when the bloom is done.

The study of pollinator decline is also interesting to some scientists, as bees have the potential to become a keystone indicator species of environmental degradation. Any changes in their abundance and diversity will influence the abundance and diversity of the prevailing plant species. This is a mutual dependency as bees rely on a steady nectar source and pollen source throughout the year to build up their hive.

### *Consequences*

The value of bee pollination in human nutrition and food for wildlife is immense and difficult to quantify. It is commonly said that about one third of human nutrition is due to bee pollination. This includes the majority of fruits, many vegetables (or their seed crop) and secondary effects from legumes such as alfalfa and clover fed to livestock.

In 2000, Drs. Roger Morse and Nicholas Calderone of Cornell University, attempted to quantify the effects of just one pollinator, the Western honey bee, on only US food crops. Their calculations came up with a figure of US $14.6 billion in food crop value. There has not been sufficient study to quantify the effects of pollinator decline on wild plants and wild life that depend on them for feed. Some plants on the endangered species list are endangered because they have lost their normal, native pollinators because of displacement by invasive honey bees. Honey bees are not native to the Western Hemisphere, so the loss of honey bees may not represent a threat to native plants;

the role of honey bees in the Western Hemisphere is almost exclusively agricultural. To the extent that honey bees compete with native bee species, a decrease in the honey bee population may be beneficial to native plants and pollinators.

### *Increasing Public Awareness*

The steady increase in beekeeper migration (for pollination service on agricultural crops) has masked the issue of pollinator decline from much public awareness, however sudden blocks to such migration could have catastrophic results on the global food supply. There are international iniatives (e.g. the International Pollinator Iniative (IPI)) that highlight the need for public participation and awareness of pollinator, such as bees, conservation.

### *Pesticide Misuse*

It is a label violation to apply most insecticides on crops during bloom, or to allow the pesticide to drift to blooming weeds that bees are visiting. Yet such applications are frequently done, with little enforcement of the bee protection directions. Pesticide misuse has driven beekeepers out of business, but can affect native wild bees even more, because they have no human to move or protect them. Bumblebee populations are in jeopardy in cotton-growing areas, since they are dosed repeatedly when pesticide applicators apply insecticides on blooming cotton fields while the bees are foraging.

Widespread aerial applications for mosquitoes, med-flies, grasshoppers, gypsy moths and other insects leave no islands of safety where wild insect pollinators can reproduce and repopulate. One such program can reduce or endanger pollinator populations for several years.

Many homeowners feel that dandelions and clover are weeds, that lawns should only be grass, and that they should be highly treated with pesticides. This makes a hostile environment for bees, butterflies and other pollinators

### *Rapid Transfer of Parasites and Diseases of Pollinator Species Around the World*

Increased international commerce within modern times has moved diseases such as American foulbrood and chalkbrood, and parasites such as varroa mites, acarina mites, and the small African hive beetle to new areas of the world, causing much loss of bees in the areas where they do not have much resistance to these pests. Imported fire ants have decimated ground nesting bees in wide areas of the southern US.

### *Loss of Habitat and Forage*

The push to remove hedgerows and other "unproductive" land in some farm areas removes habitat and homes for wild bees. Large tractor mounted rotary mowers may make farms and roadsides look neater, but they remove bee habitat at the same time. Old crops such as sweet clover and buckwheat, which were very good for bees have been disappearing. Urban and suburban development pave or build over former areas of pollinator habitat. Clearcut logging, especially when mixed forests are replaced by uniform age pine planting, causes serious loss of pollinators, by removing hardwood bloom that feeds bees early in the season, and by removing hollow trees used by feral honey bees, and dead stubs used by many solitary bees.

### *Nectar Corridors*

Migratory pollinators require a continuous supply of nectar sources to gain their energy requirements for the migration. In some areas development or agriculture has disrupted and broken up these traditional corridors, and the pollinators have to find alternative routes or discontinue migration. A good example is the endangered lesser long-nosed bat (*Leptonycteris curasoae*) which was formerly the main pollinator of a number of cactus species in southwestern United States. Its numbers have severely declined, in part due to disruption of the nectar corridors that it formerly followed. Other migratory pollinators include monarch butterflies and some hummingbirds.

### *Hive Destruction*

Bees are often viewed negatively by homeowners and other property owners. A search for "carpenter bees" on the Internet primarily yields information on removal rather than information regarding bees in a positive light. Recent hysteria regarding killer bees has contributed to these views. Beekeepers find increased vandalism of their hives, more difficulty in finding locations for bee yards, and more people inclined to sue the local beekeeper if they are stung, even if it is by a yellow jacket.

### *Light Pollution*

Increasing use of outside artificial lights, which interfere with the navigational ability of many moth species, and is suspected of interference with migratory birds may also impact pollination.

Moths are important pollinators of night blooming flowers and moth disorientation may reduce or eliminate the plants ability to reproduce, thus leading to long term ecological effects. This is a new field and this environmental issue needs further study.

### *Threat by Invasive Honey Bees*

Many native pollinators decline in population when faced with competition from invasive honey bees. For example, the western honey bee is invasive in the United States, the wild population consisting entirely of feral bees escaped from European bee colonies imported to fertilize non-native, old-world crops. Where colony collapse disorder reduced invasive honey bee populations in the US, native pollinators sometimes have made recoveries, restored to their natural niche by the loss.

### *Air Pollution*

Researchers at the University of Virginia have discovered that air pollution from automobiles and power plants has been inhibiting the ability of pollinators such as bees and butterflies to find the fragrances of flowers. Pollutants such as ozone, hydroxyl, and nitrate radicals bond quickly with volatile scent molecules of flowers, which consequently travel shorter distances intact. There results a vicious cycle in which pollinators travel increasingly longer distances to find flowers providing them nectar, and flowers receive inadequate pollination to reproduce and diversify.

### *Solutions*

The decline of pollinators is compensated to some extent by beekeepers becoming migratory, following the bloom northward in the spring from southern wintering locations. Migration may be for traditional honey crops, but increasingly is for contract pollination to supply the needs for growers of crops that require it.

### *Conservation and Restoration Efforts*

Efforts are being made to sustain pollinator diversity in agro- and natural eco-systems by some environmental groups. Prairie restoration, establishment of wildlife preserves, and encouragement of diverse wildlife landscaping rather than monoculture lawns, are examples of ways to help pollinators.

### *Use of Alternative Pollinators*

Honey bees are usually the most widely chosen insects in most managed pollination situations. However they are not the most efficient pollinators of some flowers. Alternative pollinators, such as for example, leafcutter and alkali bees in alfalfa pollination and bumblebees in greenhouses for tomatoes are used to augment and in some cases replace honey bees. A wide variety of other bees can be found in the environments that are specialist pollinators (some only using one plant species). However, most of these alternative insects' value as pollinators and their relationships with plants are as yet little known.

In the US, some think that other pollinators will in time replace the lost honey bees, blamed on introduced acarine and varroa mites, but general pollinator decline was already happening before these entered the picture. Only in a few areas are wild populations of pollinators building up; in most areas they are declining as quickly as honey bees. Furthermore pollinators cannot be exchanged on a one-for-one basis. They are not all equal. Some are generalists, some are specialists. Some are brawny; some are feeble. Some have long tongues; some short. Some work at colder temperatures than others. Bees may deliberately collect pollen, but have different collection techniques, which can greatly affect their efficiency as pollinators.

Flowers are frequently specifically adapted to one pollinator, or a small group of pollinators because of floral structure, colour, odor, nectar guides, etc. Proposed alternative pollinators may not be physically capable of accomplishing pollination, or they may not be attracted to the flower of that plant species, or they may rob nectar by cutting sepals, thus avoiding pollination. Understanding the pollination needs of a species is vital to understanding of a plant species, yet this is often poorly understood. In horticulture it is critical to the economic success of the grower, and crops have sometimes been abandoned from general use in an area because of lack of understanding of pollinator needs.

### *Importance of Pollination Management*

The increasing size of fields and orchards (monoculture) increase the importance of pollination management. Monoculture can cause a brief period when pollinators have more food resources than they can use, while other periods of the year can bring starvation or pesticide contamination of food sources. Most pollinator species rely on a steady nectar source and pollen source throughout the growing season to build up their numbers.

Crops that traditionally have had managed pollination include apple, almonds, pears, some plum and cherry varieties, blueberries, cranberries, cucumbers, cantaloupe, watermelon, alfalfa seeds, onion seeds, and many others. Some crops that have traditionally depended entirely on chance pollination by wild pollinators need pollination management nowadays to make a profitable crop.

Some crops, especially when planted in a monoculture situation, require a very high level of pollinators to produce economically viable crops. This may be because of lack of attractiveness of the blossoms, or from trying to pollinate with an alternative when the native pollinator is extinct or rare. These include crops such as alfalfa, cranberries, and kiwifruit. This technique is known as saturation pollination. In many such cases, various native bees are vastly more efficient at pollination (e.g., with blueberries), but the inefficiency of the honey bees is compensated for by using large numbers of hives, the total number of foragers thereby far exceeding the local abundance of native pollinators. In a very few cases, it has been possible to develop commercially viable pollination techniques that use the more efficient pollinators, rather than continued reliance on honey bees, as in the management of the alfalfa leafcutter bee.

It is estimated that about one hive per acre will sufficiently pollinate watermelons. In the 1950s when the woods were full of wild bee trees, and beehives were normally kept on most South Carolina farms, a farmer who grew ten acres (4 ha) of watermelons would be a large grower and probably had all the pollination needed. But today's grower may grow 200 acres (80 ha), and, if lucky, there might be one bee tree left within range. The only option in the current economy is to bring beehives to the field during blossom time.

### *Types of Pollinators*

Organisms that are currently being used as pollinators in managed pollination are honey bees, bumblebees, alfalfa leafcutter bees, and orchard mason bees. Other species are expected to be added to this list as this field develops. Humans also can be pollinators, as the gardener who hand pollinates her squash blossoms, or the Middle Eastern farmer, who climbs his date palms to pollinate them.

The Cooperative extension service recommends one honey bee hive per acre (2.5 hives per hectare) for standard watermelon varieties to meet this crop's pollination needs. In the past, when fields were small, pollination was accomplished by a mix of bees kept on farms, bumblebees, carpenter bees, feral honey bees in hollow trees and other insects.

Today, with melons planted in large tracts, the grower may no longer have hives on the farm; he may have poisoned many of the pollinators by spraying blooming cotton; he may have logged off the woods, removing hollow trees that provided homes for bees, and pushed out the hedgerows that were home for solitary native bees and other pollinating insects.

### *Planning for Improved Pollination*

Before pollination needs were understood, orchardists often planted entire blocks of apples of a single variety. Because apples are self sterile, and different members of a single variety are genetic clones (equivalent to a single plant), this is not a good idea. Growers now supply pollenizers, by planting crab apples interspersed in the rows, or by grafting crab apple limbs on some trees. Pollenizers can also be supplied by putting drum bouquets of crab apples or a compatible apple variety in the orchard blocks.

The field of pollination management cannot be placed wholly within any other field, because it bridges several fields. It draws from horticulture, apiculture, zoology (especially entomology), ecology, and botany.

## Sheet Mulching

In permaculture, sheet mulching is an agricultural no-dig gardening technique that attempts to mimic natural forests' processes. When deployed properly and in combination with other permacultural principles, it can generate healthy, productive, and low maintenance ecosystems.

### *Technique*

A model for sheet mulching consists of the following steps:

1. The area of interest is flattened by trimming down existing plant species such as grasses.
2. The soil is analysed and its pH is adjusted (if needed).
3. The soil is moisturized (if needed) to facilitate the activity of decomposers.
4. The soil is then covered with a thin layer of slowly decomposing material (known as the weed barrier), typically cardboard. This suppresses the weeds by blocking sunlight, adds nutrients to the soil as weed matter quickly decays beneath the barrier, and increases the mechanical stability of the growing medium.
5. A layer (around 10 cm thick) of weed-free soil rich in nutrients is added, in an attempt to mimic the A horizon.
6. A layer (at most 15 cm thick) of weed-free, woody and leafy matter is added in an attempt to mimic the forest floor or O Horizon. Theoretically, the soil is now ready to receive the desirable plant seeds.

Variations and considerations:

- Often the barrier is applied a few months before planting to ensure the penetration of roots of newly planted seeds.
- Very thick barriers can cause anaerobic conditions.
- Some permaculturists incorporate composting in steps 5 and/or 6.
- Sheets of newspaper and clothing can be used instead of cardboard.
- Before step 4, an initial layer (2-3 $kg/m^2$) of matter rich in nutrients (such as compost or manure) may be added to bolster decomposition.

- Some varieties of grasses and weeds may be beneficial in a number of ways. Such plants can be controlled and used rather than eradicated.
- One variation of mulching, called hugelkultur, involves using buried logs and branches as the first layer of the bed.

Advantages:

Sheet mulch has important advantages relative to conventional methods:

- Improvement of desirable plants' health and productivity.
- Retention of water and nutrients and stabilization of biochemical cycles.
- Improvement of soil structure, soil life, and prevention of soil erosion.
- Avoidance of potentially dangerous pesticides, especially herbicides.
- Reduction of overall maintenance labour and costs.

Disadvantages:

- Some weed seeds (such as those of Bermuda grass and species of bindweed) may persist under the barrier and within the soil seed bank.
- Slug populations may increase during the early stages of decomposition. However they can be kept away or harvested.
- The system may need a constant supply of organic material, at least during the early stages.
- Roaming animals may interrupt the sheet mulching process.

## Square Foot Gardening

Square foot gardening is the practice of planning and creating small but intensively planted gardens. The practice combines concepts from other organic gardening methods, including a strong focus on compost, densely planted raised beds and biointensive attention to a small, clearly defined area. Proponents claim that the method is particularly well-suited for areas with poor soil, beginner gardeners or as adaptive recreation for those with disabilities. The phrase "square foot gardening" was popularized by Mel Bartholomew in a 1981 Rodale Press book and subsequent PBS television series.

Conventional gardening can require heavy tools to loosen the soil, whereas in square foot gardening methods the soil is typically not walked on and thus not compacted, and it remains loose and more easily workable. Weeds may be easier to remove due to the light soil, and accessing them can be easier as raised beds bring the soil level closer to the gardener.

Using specific soil mixtures within the beds can help to increase water-holding capacities, so that the garden needs less additional water than in systems reliant on the native soil. Water is also spared by hand-watering directly at the plant roots, so that there is very little waste and tender young plants and seedlings are preserved.

Densely planted crops can form a living mulch, and also prevent weeds from establishing or even germinating. Natural insect repellent methods such as companion planting (i.e. planting marigolds or other naturally pest-repelling plants) become more efficient in a close space, which may reduce the need to use pesticides. The large variety of crops in a small space also prevents plant diseases from spreading easily

A plywood bottom can be attached to the bottom of a box, which can then be placed on a tabletop or raised platform for those who wish to garden without bending or squatting, or to make gardening easily accessible for wheelchair, cane or walker users. Since the beds are typcially small, making covers or cages to protect plants from pests, cold, or sun is more practical than with larger gardens. The phrase "square foot gardening" was popularized by Mel Bartholomew in a 1981 Rodale Press book and subsequent PBS television series. A full-length companion DVD, "Square Foot Gardening" (2010), was released in collaboration with Patti Moreno, the "garden girl".

The original square-foot-gardening method (per Bartholomew) uses a four-sided box with no top or bottom to contain a finite amount of soil, which was divided with a grid into sections. To encourage variety of different crops over time, each square would be planted with a different kind of plant, the number of plants per square depending on an individual plant's size. A single tomato plant might take a full square, as might herbs such as oregano, basil or mint, while most strawberry plants could be planted four per square, with up to sixteen radishes per square. Tall or climbing plants such as maize or pole beans might be planted in a northern row (south in the southern hemisphere) so as not to shade other plants, and supported with lattice or netting. The logic behind using smaller beds is that they are easily adapted, and the gardener can easily reach the entire area, without stepping on and compacting the soil. In the second edition, Bartholomew suggests using a "weed barrier" beneath the box, and filling it completely with "Mel's mix," a combination by volume of one third of decayed Sphagnum peat moss, one-third expanded vermiculite and one-third blended compost. New compost should be added and mixed in each year. For accessibility, raised boxes may have bottoms to sit like tables at a convenient height, with approximately 6" (15 cm) of manufactured soil per square foot. For some plants, such as carrots or asparagus, it is recommended to build areas deeper than 6" in order to facilitate a deeper root requirement.

In Bartholomew's method, the garden space is divided into beds that are easily accessed from every side. A 4' x 4', 16 sq ft (1.5 $m^2$) or 120 cm x 120 cm, 1.4 $m^2$ garden is recommended for the first garden, and a path wide enough to comfortably work from should be made on each side of the bed, if possible, or if the bed must be accessed by reaching across it, a narrower one should be used so that no discomfort results from tending the garden. Each of the beds is divided into approximately one square foot units and marked out with sticks, twine, or sturdy slats to ensure that the square foot units remain visible as the garden matures.

Different seeds are planted in each square, to ensure a rational amount of each type of crop is grown, and to conserve seeds instead of overplanting, crowding and thinning plants. Common spacing is one plant per square for larger plants (broccoli, basil, etc.), four plants per square for medium large plants like lettuce, nine plants per square for medium-small plants like spinach, and sixteen per square for small plants such as onions and carrots. Plants that normally take up yards of space as runners, such as squash or cucumbers, are grown vertically on sturdy frames that are hung with netting or string to support the developing crops. Ones that grow deep underground, such as potatoes or carrots, are grown in a square foot section that has foot tall sides and a planting surface above the ground, so that a foot or more of framed soil depth is provided above the garden surface rather than below it.

The beds are weeded and watered from the pathways, so the garden soil is never stepped on or compacted. Because a new soil mixture is used to create the garden, and a few handfuls of compost are added with each harvest to maintain soil fertility over time, the state of the site's underlying soil is irrelevant. This gardening method has been employed successfully in every region, including in deserts, on high arid mountain plateaus, in cramped urban locations, and in areas with polluted or high salinity soils. It is equally useful for growing flowers, vegetables, herbs and some fruits in containers, raised beds, on tabletops or at ground level, in only 4 to 6 inches (150 mm) of soil. A few seeds per square foot, the ability to make compost, to water by hand, and to set up the initial garden in a sunny position or where a container, table or platform garden may be moved on wheels to receive light is all that is needed to set up a square foot garden.

# 5

# Succession Planting

In agriculture, succession planting refers to several planting methods that increase crop availability during a growing season by making efficient use of space and timing.

There are four basic approaches, that can also be combined:

- Two or more crops in succession: After one crop is harvested, another is planted in the same space. The length of the growing season, climate, and crop selection are key factors. For example, a cool season spring crop could be followed by a heat-loving summer crop.
- Same crop, successive plantings: Several smaller plantings are made at timed intervals, rather than all at once. The plants mature at staggered dates, establishing a continuous harvest over an extended period. Lettuce and other salad greens are common crops for this approach.
- Two or more crops simultaneously: Non-competing crops, often with different maturity dates, are planted together in various patterns. Intercropping is one pattern approach; companion planting is a related, complementary practice.
- Same crop, different maturity dates: Several varieties are selected, with different maturity dates: early, main season, late. Planted at the same time, the varieties mature one after the other over the season.

These techniques can be used to design complex, highly productive cropping systems. The more involved the plan, the more detailed knowledge is required, of the specific varieties and how they perform in a particular growing location. The term "succession planting" usually appears in literature for home gardening and small-scale farming, although the techniques apply to any scale. Some definitions include one or more, but not all of the four techniques described above.

Succession planting is often used in organic farming. Multiple cropping describes essentially the same general method. A catch crop refers to a specific type of succession planting, where a fast-growing crop is grown simultaneously with, or between successive plantings of, a main crop.

## Weed Control

Weed control is the botanical component of pest control, using physical and chemical methods to stop weeds from reaching a mature stage of growth when they could be harmful to domesticated plants and livestock. In order to reduce weed growth, many "weed control" strategies have been developed in order to contain the growth and spread of weeds. The most basic is ploughing which cuts the roots of annual weeds. Today, chemical weed killers known as herbicides are widely used.

### *Exact Definition*

There is no universal definition for what qualifies as an obnoxious plant. However, a plant is often termed weed when it has one or more of the following characteristics:

- Little or no value (as in medicinal, nutritional, or energy)
- Very high growth rate and/or ease of germination
- Exhibits competition to crops, for space, light, water and nutrients

### *Effects on Other Plants*

Weeds can compete with productive crops or pasture, or convert productive land into unusable scrub. Weeds are also often poisonous, distasteful, produce burrs, thorns or other damaging body parts or otherwise interfere with the use and management of desirable plants by contaminating harvests or excluding livestock.

Weeds tend to thrive at the expense of the more refined edible or ornamental crops. They provide competition for space, nutrients, water and light, although how seriously they will affect a crop depends on a number of factors. Some crops have greater resistance than others- smaller, slower growing seedlings are more likely to be overwhelmed than those that are larger and more vigourous. Onions are one of the crops most susceptible to competition, for they are slow to germinate and produce slender, upright stems. Quick growing, broad leafed weeds therefore have a distinct advantage, and if not removed, the crop is likely to be lost. Broad beans however produce large seedlings, and will suffer far less profound effects of weed competition other than during periods of water shortage at the crucial time when the pods are filling out. Transplanted crops raised in sterile seed or potting compost will have a head start over germinating weed seeds.

Weeds also differ in their competitive abilities, and can vary according to conditions and the time of year. Tall growing vigourous weeds such as fat hen (*Chenopodium album*) can have the most pronounced effects on adjacent crops, although seedlings of fat hen that appear in late summer will only produce small plants. Chickweed (*Stellaria media*), a low growing plant, can happily co-exist with a tall crop during the summer, but plants that have overwintered will grow rapidly in early spring and may swamp crops such as onions or spring greens.

The presence of weeds does not necessarily mean that they are competing with a crop, especially during the early stages of growth when each plant can find the resources it requires without interfering with the others. However, as the seedlings' size increases, their root systems will spread as they each begin to require greater amounts of water and nutrients. Estimates suggest that weed and crop can co-exist harmoniously for around three weeks, therefore it is important that weeds be removed early on in order to prevent competition occurring. Weed competition can have quite dramatic effects on crop growth. Harold A Roberts cites research carried out with onions wherein *"Weeds were carefully removed from separate plots at different times during the growth of the crop and the plots were then kept clean. It was found that after competition had started, the final yield of bulbs was being reduced at a rate equivalent to almost 4% per day. So that by delaying weeding for another fortnight, the yield was cut to less than half that produced on ground kept clean all the time."* (*The Complete Know And Grow Vegetables*,

Bleasdale, Salter and others, OUP 1991). He goes on to record that *"by early June, the weight of weeds per unit area was twenty times that of the crop, and the weeds had already taken from the soil about half of the nitrogen and a third of the potash which had been applied"*.

Perennial weeds with bulbils, such as lesser celandine and oxalis, or with persistent underground stems such as couch grass (*Agropyron repens*) or creeping buttercup (*Ranunculus repens*) are able to store reserves of food, and are thus able to grow faster and with more vigour than their annual counterparts. There is also evidence that the roots of some perennials such as couch grass exude allelopathic chemicals which inhibit the growth of other nearby plants.

Weeds can also host pests and diseases that can spread to cultivated crops. Charlock and Shepherd's purse may carry clubroot, eelworm can be harboured by chickweed, fat hen and shepherd's purse, while the cucumber mosaic virus, which can devastate the cucurbit family, is carried by a range of different weeds including chickweed and groundsel.

However, at times the role of weeds in this respect can be over-rated. As far as insect pests are concerned, often the species that live on weeds are not the same as those that attack vegetable crops; *"Tests with the common cruciferous weeds such as shepherds purse have shown that they do not act as hosts for the larvae of the cabbage root fly. One exception was found to be the wild radish, but this is not usually a weed of established vegetable gardens"* (Roberts, *The Complete Know And Grow Vegetables*). However pests such as cutworms may first attack weeds then move on to cultivated crops.

While charlock, a common weed in southeastern USA, may be considered a weed by row crop growers, it is highly valued by beekeepers, who seek out places where it blooms all winter, thus providing pollen for honeybees and other pollinators. Its bloom is resistant to all but a very hard freeze, and even that will only kill it back briefly. By feeding an array of pollinators during a seasonal dearth, it can redound to the farmer's advantage. Many weeds are likewise highly beneficial to pollinators.

In domestic gardens, methods of weed control include covering an area of ground with several layers of wet newspaper or one black plastic sheet for several weeks. In the case of using wet newspaper, the multiple layers prevent light from reaching all plants beneath, which kills them. Saturating the newspaper with water daily speeds the decomposition of the dead plants. Any weed seeds that start to sprout because of the water will also be deprived of sunlight, be killed, and decompose. After several weeks, all germinating weed seeds present in the ground should be dead. Then the newspaper can be removed and the ground can be planted. The decomposed plants will help fertilise the plants or seeds planted later.

In the case of using the black plastic sheet, the greenhouse effect is used to kill the plants beneath the sheet. A 5-10 cm layer of wood chip mulch on the ground will also prevent most weeds from sprouting. Also, gravel can be spread over the ground as an inorganic mulch. In agriculture, irrigation is sometimes used as a weed control measure such as in the case of paddy fields. Many people find that although the black plastic sheeting is extremely effective at preventing the weeds in areas where it covers, but in actual use it is difficult to achieve full coverage.

Knowing how weeds reproduce, spread and survive adverse conditions can help in developing effective control and management strategies. Weeds have a range of techniques that enable them to thrive; Annual and biennial weeds such as chickweed, annual meadow grass, shepherd's purse, groundsel, fat hen, cleaver, speedwell and hairy bittercress propagate themselves by seeding. Many produce huge numbers of seed several times a season, some all year round. Groundsel can produce 1000 seed, and can continue right through a mild winter, whilst Scentless Mayweed produces over 30,000 seeds per plant. Not all of these will germinate at once, but over several seasons, lying dormant in the soil sometimes for years until exposed to light. Poppy seed can survive 80-100 years, dock 50 or more. There can be many thousands of seeds in a square foot or square metre of ground, thus and soil disturbance will produce a flush of fresh weed seedlings.

### *"Stale Seed bed" Technique*

One technique employed by growers is the 'stale seed bed', which involves cultivating the soil, then leaving it for a week or so. When the initial flush of weeds has germinated, the grower will lightly hoe off before the desired crop is planted. However, even a freshly cleared bed will be susceptible to airborne seed from elsewhere, as well as seed brought in by passing animals which can carry them on their fur, or from freshly imported manure. The organic solution to the problem of spreading annual weeds lies in regular, properly timed weeding, preferably just before flowering (fortuitously, this is also the time at which they will be of the most value in composting).

This technique is also quite often used by farmers who let weeds germinate then return the soil before crop sowing.

Perennial weeds also propagate by seeding; the airborne seed of the dandelion and the rose-bay willow herb are parachuted far and wide. But they also have an additional range of vegetative means of spreading that gives them their pernicious reputation. Dandelion and dock put down deep tap roots, which, although they do not spread underground, are able to regrow from any remaining piece left in the ground. Removal of the complete tap root is the only sure remedy.

The most persistent of the perennials are those that spread by underground creeping rhizomes that can regrow from the tiniest fragment.

These include couch grass, bindweed, ground elder, nettles, rosebay willow herb, Japanese knotweed, horsetail and bracken, as well as creeping thistle, whose tap roots can put out lateral roots. Other perennials put out runners that spread along the soil surface. As they creep along they set down roots, enabling them to colonise bare ground with great rapidity. These include creeping buttercup and ground ivy. Yet another group of perennials propagate by stolons- stems that arch back into the ground to reroot. Most familiar of these is the bramble.

All of the above weeds can be very difficult to eradicate- thick black plastic mulches can be effective to a degree, although will probably need to be left in place for at least two seasons. In addition, hoeing off weed leaves and stems as soon as they appear can eventually weaken and kill the plants, although this will require persistence in the case of plants such as bindweed. Nettle infestations can be tackled by cutting back at least three times a year, repeated over a three year period. Bramble can be dealt with in a similar way. Some plants are said to produce root exudates that suppress herbaceous weeds. *Tagetes minuata* is claimed to be effective against couch and ground elder, whilst a border of comfrey is also said to act as a barrier against the invasion of some weeds including couch.

### *Use of Herbicides*

The above described methods of weed control avoid using chemicals. They are often used by farmers. However, these methods may damage a fragile soil by restructuring it, hence are not always used. They are those preferred by the organic gardener or organic farmer.

However weed control can also be achieved by the use of herbicides. Selective herbicides kill certain targets while leaving the desired crop relatively unharmed. Some of these act by interfering with the growth of the weed and are often based on plant hormones. Herbicides are generally classified as follows;

- Contact herbicides destroy only that plant tissue in contact with the chemical spray. Generally, these are the fastest acting herbicides. They are ineffective on perennial plants that are able to re-grow from roots or tubers.
- Systemic herbicides are foliar-applied and are translocated through the plant and destroy a greater amount of the plant tissue. Modern herbicides such as glyphosate are designed to leave no harmful residue in the soil.
- Soil-borne herbicides are applied to the soil and are taken up by the roots of the target plant.
- Pre-emergent herbicides are applied to the soil and prevent germination or early growth of weed seeds.

In agriculture large scale and systematic weeding is usually required, often by machines, such as liquid herbicide sprayers, or even by helicopter (such as in the USA), to eliminate the massive amount of weeds present on farming lands. However there are a number of techniques that the organic farmer can employ such as mulching and carefully timed cutting of weeds before they are able to set seed.

### *Organic Methods*

Typically combinations of methods are used in organic situations.

- Drip irrigation: Rubber hoses and other methods are used to bring water directly to the roots of the desired plants. This limits weed access to water.
- Manually pulling weeds: Labourers are used to pull weeds at various points in the growing process.
- Boiling water: Pour boiling water to weed, they will become more green and then die in few hours. Best for weed in cracks or other hard to reach locations.
- Vinegar: Vinegar kills the visible part of the weed. They will wrinkle and die next day, although the root will still be in place to continue growing.
- Mechanically tilling around plants: Tractors are used to carefully till weeds around the crop plants at various points in the growing process. Besides tilling, other mechanical weed control methods also exist
- Ploughing: Ploughing includes tilling of soil, intercultural ploughing and summer ploughing. Ploughing through tilling of soil uproots the weeds which causes them to die. In summer ploughing is done during deep summers. Summer ploughing also helps in killing pests.
- Crop rotation: Rotating crops with ones that kill weeds by choking them out, such as hemp, *Mucuna pruriens*, and other crops, can be a very effective method of weed control. It is a way to avoid the use of herbicides, and to gain the benefits of crop rotation.
- Weed mat: A weed mat is an artificial mulch, fibrous cloth material, bark or newspaper laid on top of the soil preventing weeds from growing to the surface.

### *Thermal Methods*

There are several thermal methods known to control weed. Weed burners heat up soil quickly and destroy superficial parts of the plants. Weed seeds are often heat resistant and even react with an increase of growth on dry heat. Since the 19th century soil steam sterilization is used as a farming technique to clean soil completely from weeds. Several research results confirm the high effectivness of humid heat against weeds and its seeds.

The *Weeds Act, 1959* is described as *"Preventing the spread of harmful or injurious weeds"*, and is mainly relevant to farmers and other rural settings rather than the allotment or garden-scale grower. There are five 'injurious' (that is, likely to be harmful to agricultural production) weeds covered by the provisions of the Weeds Act. These are:

- Spear thistle (*Cirsium vulgare*)
- Creeping or field thistle (*Cirsium arvense*)
- Curled Dock (*Rumex crispus*)
- Broad leaved dock (*Rumex obtusifolius*)
- Common ragwort (*Senecio jacobaea*) (n.b., this weed is poisonous to livestock. Livestock should not be allowed to graze where ragwort has grown until it is eradicated, and any traces have disintegrated. Ragwort should not be allowed to be harvested in hay or silage for feed).

The Department for Environment, Food and Rural Affairs (DEFRA) provides guidance for the treatment removal of these weeds from infested land. Much of this is oriented towards the use of herbicides, the majority of which may not be acceptable to the organic producer (apart from non-synthetic substances like sulphur, which in some circumstances are accepted within Soil Association standards) but in most cases there are manual techniques that can be used such as digging out the roots, mulching out or carefully timed cutting before seeds are able to spread.

Primary responsibility for weeds control rests with the occupier of the land on which the weeds are growing, therefore it is important to be alert to potential weed problems and to take prompt action. However, it should be remembered that most common farmland weeds are not "injurious" within the meaning of the Weeds Act, and many such plant species have conservation and environmental value.

When dealing with complaints under the Weeds Act, DEFRA has a duty in law to try to achieve a reasonable balance among different interests. These include agriculture, countryside conservation and the public in general. Constructive discussion about problems caused by weeds can often result in effective solutions and avoid the need for DEFRA to take official action. In addition to those weeds covered by the 1959 act, under section 14 of the *Wildlife and Countryside Act 1981*, it can be an offence to plant or grow certain specified plants in the wild, including Giant Hogweed and Japanese Knotweed.

Problems involving these plants can be referred to the local authority for the area where those weeds are growing as some local authorities have bye-laws controlling these plants. There is no statutory requirement for landowners to remove these plants from their property.

### *Invasive Species*

Invasive species, also called invasive exotics or simply exotics, is a nomenclature term and categorization phrase used for flora and fauna, and for specific restoration-preservation processes in native habitats, with several definitions.

- The first definition, the most used, applies to introduced species (also called "non-indigenous" or "non-native") that adversely affect the habitats and bioregions they invade economically, environmentally, and/or ecologically. Such invasive species may be either plants or animals and may disrupt by dominating a region, wilderness areas, particular habitats, or wildland-urban interface land from loss of natural controls (such as predators or herbivores). This includes non-native invasive plant species labelled as exotic pest plants and invasive exotics growing in native plant communities. It has been used in this sense by government organisations as well as conservation groups such as the International Union for Conservation of Nature (IUCN) and the California Native Plant Society. The European Union defines "Invasive Alien Species" as those that are, firstly, outside their natural distribution area, and secondly, threaten biological diversity. It is also used by land managers, botanists, researchers, horticulturalists, conservationists, and the public for noxious weeds. The kudzu vine (*Pueraria lobata*),

Andean Pampas grass (*Cortaderia jubata*), and yellow starthistle (*Centaurea solstitialis*) are examples.

- The second definition includes the first, but broadens the boundaries to include indigenous or *native* species, with the *non-native* ones, that disrupt by a dominant colonization of a particular habitat or wildlands area from loss of natural controls (i.e.: predators or herbivores). Deer are an example, considered to be overpopulating their native zones and adjacent suburban gardens, by some in the Northeastern and Pacific Coast regions of the United States.
- The third definition identifies invasive species as a *widespread nonindigenous species*. This one can be too broad, as not every nonindigenous or "introduced" species has an adverse effect on a nonindigenous environment. A nonadverse example is the common goldfish (*Carassius auratus*), though common outside its native range globally, it is rarely in harmful densities to a native habitat.

Because of the variability of its definition, and because definitions are often from a socioeconomic perspective, the phrase *invasive species* is often criticized as an imprecise term for the scientific field of ecology.

### *Conditions that Lead to Invasion*

Scientists include species- and ecosystem factors among the mechanisms, that when combined establish invasiveness in a newly introduced species.

### *Species-based Mechanisms*

While all species compete to survive, invasive species appear to have specific traits or specific combinations of traits that allow them to outcompete native species. In some cases the competition is about rates of growth and reproduction. In other cases species interact with each other more directly.

Researchers disagree about the usefulness of traits as invasiveness markers. One study found that of a list of invasive and noninvasive species, 86% of the invasive species could be identified from the traits alone. Another study found invasive species tended only to have a small subset of the presumed traits, and that many such traits were found in noninvasive species, requiring other explanations. Common invasive species traits include:

- Fast growth
- Rapid reproduction
- High dispersal ability
- Phenotypic plasticity (the ability to alter growth form to suit current conditions)
- Tolerance of a wide range of environmental conditions (Ecological competence)
- Ability to live off of a wide range of food types (generalist)
- Association with humans
- Prior successful invasions.

Typically an introduced species must survive at low population densities before it becomes invasive in a new location. At low population densities, it can be difficult for the introduced species to reproduce and maintain itself in a new location, so a species might reach a location multiple times before it becomes established. Repeated patterns of human movement, such as ships sailing to and from ports or cars driving up and down highways, offer repeated opportunities for establishment (also known as a high propagule pressure).

An introduced species might become invasive if it can outcompete native species for resources, such as nutrients, light, physical space, water or food. If these species evolved under great competition or predation, the new environment may host fewer able competitors, allowing the invader to proliferate quickly. Ecosystems in which all available resources are being used to their fullest capacity by native species can be modelled as zero-sum systems, where any gain for the invader is a loss for the native.

However, such unilateral competitive superiority (and extinction of native species with increased populations of the invader) is not the rule. Invasive species often coexist with native species for an extended time, and gradually the superior competitive ability of an invasive species becomes apparent as its population grows larger and denser and it adapts to its new location.

An invasive species might be able to use resources previously unavailable to native species, such as deep water sources accessed by a long taproot, or an ability to live on previously uninhabited soil types. For example, barbed goatgrass (*Aegilops triuncialis*) was introduced to California on serpentine soils, which have low water-retention, low nutrient levels, a high Magnesium/Calcium ratio, and possible heavy metal toxicity.

Plant populations on these soils tend to show low density, but goatgrass can form dense stands on these soils, crowding out native species that have not adapted well to serpentine soils. Ecological facilitation occurs when a species alters its environment using chemicals or manipulating abiotic factors, allowing the species to thrive, while making the environment less favourable to competitors. One such facilitative mechanism is allelopathy, also known as chemical competition or interference competition. In allelopathy, a plant secretes chemicals which make the surrounding soil uninhabitable, or at least inhibitory, to competing species.

Examples of this in Centaurea are *Centaurea solstitialis* (yellow starthistle) and *Centaurea diffusa* (diffuse knapweed). These Eastern European noxious weeds have spread through the western and West Coast states. Experiments show that 8-hydroxyquinoline, a chemical produced at the root of *C. diffusa*, has a negative effect only on plants that have not co-evolved with it. Such co-evolved native plants have also evolved defences. *C. diffusa* and *C. solstitialis* do not appear in their native habitats to be overwhelmingly successful competitors. Success/lack of success in one habitat does not imply success in others. Conversely, examining habitats in which a species is less successful can reveal novel weapons to defeat invasiveness.

Changes in fire regimens are another form of facilitation. *Bromus tectorum*, originally from Eurasia, is highly fire-adapted. It not only spreads rapidly after burning, but also increases the frequency and intensity (heat) of fires, by providing large amounts of dry detritus during the fire season in western North America. In areas where it is widespread, it has altered the local fire regimen so much that native plants cannot survive the frequent fires, allowing *B. tectorum* to further extend and maintain dominance in its introduced range.

Facilitation also occurs when one species physically modifies a habitat in ways that are advantageous to other species. For example, zebra mussels increase habitat complexity on lake floors, providing crevices in which invertebrates live. This increase in complexity, together with the nutrition provided by the waste products of mussel filter-feeding, increases the density and diversity of benthic invertebrate communities.

### *Ecosystem-based Mechanisms*

In ecosystems, the amount of available resources and the extent to which those resources are used by organisms determines the effects of additional species on the ecosystem. In stable ecosystems, equilibrium exists in the use of available resources. These mechanisms describe a situation in which the ecosystem has suffered a disturbance which changes the fundamental nature of the ecosystem.

When changes such as a forest fire occur, normal succession favours native grasses and forbs. An introduced species that can spread faster than natives can use resources that would have been available to native species, squeezing them out. Nitrogen and phosphorus are often the limiting factors in these situations. Every species occupies a *niche* in its native ecosystem; some species fill large and varied roles, while others are highly specialised. Some invading species fill niches that are not used by native species, and they also can create new niches.

Ecosystem changes can alter species' distributions. For example edge effects describe what happens when part of an ecosystem is disturbed as when land is cleared for agriculture. The boundary between remaining undisturbed habitat and the newly cleared land itself forms a distinct habitat, creating new winners and losers and possibly hosting species that would not thrive outside the boundary habitat.

Although invasive species have typically been introduced to a habitat, some native species can, under the influence of events, such as long-term rainfall changes or human modifications to the habitat, increase in number and range and become invasive by expanding into new areas and disturbing the balance of species in the new area.

All species experience increases and decreases in numbers, in many cases accompanied by expansion or contraction of range. For example, the Monterey cypress is an endangered endemic, naturally occurring only in two small stands in California. They are being exterminated as exotic invasive species less than 50 miles (80 km) from their natural habitat.

### *Traits of Invaded Ecosystems*

In 1958, Charles S. Elton claimed that ecosystems with higher species diversity were less subject to invasive species because of fewer available niches. Other ecologists later pointed to highly diverse, but heavily invaded ecosystems and argued that ecosystems with high species diversity were more susceptible to invasion.

This debate hinged on the spatial scale at which invasion studies were performed, and the issue of how diversity affects susceptibility remained unresolved as of 2011. Small-scale studies tended to show a negative relationship between diversity and invasion, while large-scale studies tended to show the reverse. The latter result may be a side-effect of invasives' ability to capitalize on increased resource availability and weaker species interactions that are more common when larger samples are considered.

Invasion was more likely in ecosystems that were similar to the one in which the potential invader evolved. Island ecosystems may be more prone to invasion because their species faced few strong competitors and predators, or because their distance from colonizing species populations makes them more likely to have "open" niches.

An example of this phenomenon was the decimation of native bird populations on Guam by the invasive brown tree snake. Conversely, invaded ecosystems may lack the natural competitors and predators that check invasives' growth in their native ecosystems, a factor that affected Guam snake populations. Invaded ecosystems may have experienced disturbance, typically human-induced. Such a disturbance may give invasive species a chance to establish themselves with less competition from natives less able to adapt to a disturbed ecosystem.

### *Vectors*

Non-native species have many *vectors*, including biogenic vectors, but most invasions are associated with human activity.

Natural range extensions are common in many species, but the rate and magnitude of human-mediated extensions in these species tend to be much larger than natural extensions, and humans typically carry specimens greater distances than natural forces. An early human vector occurred when prehistoric humans introduced the Pacific rat (*Rattus exulans*) to Polynesia.

Vectors include plants or seeds imported for horticulture. The pet trade moves animals across borders, where they can escape and become invasive. Organisms stow away on transport vehicles. For example, ballast water taken up at sea and released in port is a major vector for marine invasions. Freshwater zebra mussels, native to the Black, Caspian and Azov seas, probably reached the Great Lakes via ballast water from a transoceanic vessel.

The arrival of invasive propagules to a new site is a function of the site's invasibility.

Species have also been introduced intentionally. For example, to feel more "at home", American colonists formed "Acclimation Societies" that repeatedly imported birds that were native to Europe to North America and other distant lands. In 2008, U.S. postal workers in Pennsylvania noticed noises coming from inside a box from Taiwan; the box contained more than two dozen live beetles. Agricultural Research Service entomologists identified them as rhinoceros beetle, hercules beetle, and king stag beetle.

Because these species were not native to the U.S., they could have threatened native ecosystems. To prevent exotic species from becoming a problem in the U.S., special handling and permits are required when living materials are shipped from foreign countries. USDA programs such as Smuggling Interdiction and Trade Compliance (SITC) attempt to prevent exotic species outbreaks in America.

Economics plays a major role in exotic species introduction. High demand for the valuable Chinese mitten crab is one explanation for the possible intentional release of the species in foreign waters.

### *Impacts of Wildfire*

Invasive species often exploit disturbances to an ecosystem (wildfires, roads, foot trails) to colonize an area. Large wildfires are capable of sterilizing soils, while adding a variety of nutrients. In the resulting free-for-all, formerly entrenched species lose their advantage, leaving more room for invasives. In such circumstances plants that can regenerate from their roots have an advantage.

Non-natives with this ability can benefit from a low intensity fire burns that removes surface vegetation, leaving natives that rely on seeds for propagation to find their niches occupied when their seeds finally sprout.

### *Impact of Wildfire Suppression on Spreading*

Wildfires often occur in remote areas, needing fire suppression crews to travel through pristine forest to reach the site. The crews can bring invasive seeds with them. If any of these stowaway seeds become established, a thriving colony of invasives can erupt in as few as six weeks, after which controlling the outbreak can need years of continued attention to prevent further spread.

Also, disturbing the soil surface, such as cutting firebreaks, destroys native cover, exposes soil, and can accelerate invasions. In suburban and wildland-urban interface areas, the vegetation clearance and brush removal ordinances of municipalities for defensible space can result in excessive removal of native shrubs and perennials that exposes the soil to more light and less competition for invasive plant species.

Fire suppression vehicles are often major culprits in such outbreaks, as the vehicles are often driven on back roads often overgrown with invasive plant species. The undercarriage of the vehicle becomes a prime vessel of transport. In response, on large fires, washing stations "decontaminate" vehicles before engaging in suppression activities. Large wild fires attract firefighters from remote locales, further increasing the potential for seed transport.

Land clearing and human habitation put significant pressure on local species. Disturbed habitats are prone to invasions that can have adverse effects on local ecosystems, changing ecosystem functions. A species of wetland plant known as aeae in Hawaii (the indigenous *Bacopa monnieri*) is regarded as a pest species in artificially manipulated water bird refuges because it quickly covers shallow mudflats established for endangered Hawaiian stilt (*Himantopus mexicanus knudseni*), making these undesirable feeding areas for the birds.

Multiple successive introductions of different non-native species can have interactive effects; the introduction of a second non-native species can enable the first invasive species to flourish. Examples of this are the introductions of the amethyst gem clam (*Gemma gemma*) and the European green crab (*Carcinus maenas*).

The gem clam was introduced into California's Bodega Harbor from the East Coast of the United States a century ago. It had been found in small quantities in the harbor but had never displaced the native clam species (*Nutricola* spp.). In the mid 1990s, the introduction of the European green crab, found to prey preferentially on the native clams, resulted in a decline of the native clams and an increase of the introduced clam populations.

In the Waterberg region of South Africa, cattle grazing over the past six centuries has allowed invasive scrub and small trees to displace much of the original grassland, resulting in a massive reduction in forage for native bovids and other grazers. Since the 1970s, large scale efforts have been underway to reduce invasive species; partial success has led to re-establishment of many species that had dwindled or left the region. Examples of these species are giraffe, blue wildebeest, impala, kudu and white rhino.

Invasive species can change the functions of ecosystems. For example, invasive plants can alter the fire regimen (cheatgrass, *Bromus tectorum*), nutrient cycling (smooth cordgrass *Spartina alterniflora*), and hydrology (*Tamarix*) in native ecosystems. Invasive species that are closely related to rare native species have the potential to hybridize with the native species. Harmful effects of hybridization have led to a decline and even extinction of native species. For example, hybridization with introduced cordgrass, *Spartina alterniflora*, threatens the existence of California cordgrass (*Spartina foliosa*) in San Francisco Bay.

Non-native species can have benefits. Asian oysters, for example, better filter water pollutants than native oysters.

They also grow faster and withstand disease better than natives. Biologists are currently considering releasing this mollusk in the Chesapeake Bay to help restore oyster stocks and remove pollution. A recent study by the Johns Hopkins School of Public Health found the Asian oyster could significantly benefit the bay's deteriorating water quality. Economic costs from invasive species can be separated into direct costs through production loss in agriculture and forestry, and management costs. Estimated damage and control cost of invasive species in the U.S. alone amount to more than $138 billion annually. Economic losses can also occur through loss of recreational and tourism revenues.

When economic costs of invasions are calculated as production loss and management costs, they are low because they do not consider environmental damage; if monetary values were assigned to the extinction of species, loss in biodiversity, and loss of ecosystem services, costs from impacts of invasive species would drastically increase. The following examples from different sectors of the economy demonstrate the impact of biological invasions.

Some invasions offer potential commercial benefits. For instance, silver carp and common carp can be harvested for human food and exported to markets already familiar with the product, or processed into pet foods, or mink feed. Vegetative invasives such as water hyacinth can be turned into fuel by methane digesters.

Weeds reduce yield, though they may provide essential nutrients. Some deep-rooted weeds can "mine" nutrients from the subsoil and deposit them on the topsoil, while others provide habitat for beneficial insects and/or provide foods for pest species. Many weed species are accidental introductions that accompany seeds and imported plant material. Many introduced weeds in pastures compete with native forage plants, threaten young cattle (e.g., leafy spurge, *Euphorbia esula*) or are unpalatable because of thorns and spines (e.g., yellow starthistle).

Forage loss from invasive weeds on pastures amounts to nearly US$1 billion in the U.S. alone. A decline in pollinator services and loss of fruit production has been caused by honey bees infected by the invasive varroa mite. Introduced rats (*Rattus*

*rattus* and *R. norvegicus*) have become serious pests on farms, destroying stored grains.

The unintentional introduction of forest pest species and plant pathogens can change forest ecology and damage the timber industry. The Asian long-horned beetle (*Anoplophora glabripennis*) was first introduced into the U.S. in 1996, and was expected to infect and damage millions of acres of hardwood trees. As of 2005 thirty million dollars had been spent in attempts to eradicate this pest and protect millions of trees in the affected regions.

In North America, the woolly adelgid has inflicted damage on old-growth spruce fir and hemlock forests and damages the Christmas tree industry. The chestnut blight fungus (*Cryphonectria parasitica*) and Dutch elm disease (*Ophiostoma novo-ulmi*) are two plant pathogens with serious impacts on these two species, and forest health.

Invasive species can impact outdoor recreation, such as fishing, hunting, hiking, wildlife viewing, and water-based activities. They can damage a wide array of environmental services that are important to recreation, including, but not limited to, water quality and quantity, plant and animal diversity, and species abundance.

Eiswerth states, "very little research has been performed to estimate the corresponding economic losses at spatial scales such as regions, states, and watersheds." Eurasian Watermilfoil (*Myriophyllum spicatum*) in parts of the US, fill lakes with plants complicating fishing and boating.

Encroachment of humans into previously remote ecosystems has exposed exotic diseases such as AIDS virus to the wider population. Introduced birds (e.g. pigeons), rodents and insects (e.g. mosquito, flea, louse and tsetse fly pests) can serve as vectors and reservoirs of human afflictions. The introduced Chinese mitten crabs are carriers of Asian lung fluke. Throughout recorded history, epidemics of human diseases, such as malaria, yellow fever, typhus, and bubonic plague, spread via these vectors.

A recent example of an introduced disease is the spread of the West Nile virus, which killed humans, birds, mammals, and reptiles. Waterborne disease agents, such as cholera bacteria

(*Vibrio cholerae*), and causative agents of harmful algal blooms are often transported via ballast water. Invasive species and accompanying control efforts can have long term public health implications. For instance, pesticides applied to treat a particular pest species could pollute soil and surface water.

Biotic invasion is considered one of the five top drivers for global biodiversity loss and is increasing because of tourism and globalization. This may be particularly true in inadequately regulated fresh water systems, though quarantines and ballast water rules have improved the situation. Invasive species may drive local native species to extinction via competitive exclusion, niche displacement, or hybridisation with related native species.

Therefore, besides their economic ramifications, alien invasions may result in extensive changes in the structure, composition and global distribution of the biota of sites of introduction, leading ultimately to the homogenisation of the world's fauna and flora and the loss of biodiversity.

Nevertheless, it is difficult to unequivocally attribute extinctions to a species invasion, and the few scientific studies that have done so have been with animal taxa. Concern over the impacts of invasive species on biodiversity must therefore consider the actual evidence (either ecological or economic), in relation to the potential risk.

Native species can be threatened with extinction through the process of *genetic pollution*. Genetic pollution is unintentional hybridization and introgression, which leads to homogenization or replacement of local genotypes as a result of either a numerical or fitness advantage of the introduced species. Genetic pollution can operate either through introduction or through habitat modification, bringing previously isolated species into contact.

Hybrids resulting from rare species that interbreed with abundant species can swamp the rarer species' gene pool. This is not always apparent from morphological observations alone. Some degree of gene flow is normal, and preserves constellations of genes and genotypes. An example of this is the interbreeding of migrating coyotes with the red wolf, in areas of eastern North Carolina where the red wolf was reintroduced.

### *Biogeographic Evaluation*

| *Stage* | *Characteristic* |
|---|---|
| 0 | Propagules residing in a donor region |
| I | Travelling |
| II | Introduced |
| III | Localized and numerically rare |
| IVa | Widespread but rare |
| IVb | Localized but dominant |
| V | Widespread and dominant |

In an attempt to avoid the ambiguous, subjective, and pejorative vocabulary that so often accompanies discussion of invasive species even in scientific papers, Colautti and MacIsaac proposed a new nomenclature system based on biogeography rather than on taxa.

By discarding taxonomy, human health, and economic factors, this model focused only on ecological factors. The model evaluated individual populations rather than entire species. It classified each population based on its success in that environment. This model applied equally to indigenous and to introduced species, and did not automatically categorize successful introductions as harmful.

# 6

# Motivations for Organic Agriculture

Within the food industry, defining the benefits of organic food is largely left to word of mouth, media coverage, and the promotional efforts of organic advocates. Major food and beverage corporations have rapidly moved to acquire significant stake in both fresh and processed organic products.

Still, the specific sales points of "organics" go largely unmentioned on product packaging and in mainstream media advertising. Claims of improved food quality are regularly used in conventional food marketing, with "low fat", "low sodium", "whole grain", "high fibre", "vitamin enriched", "no trans-fat" and other commonly advertised benefits.

By contrast, "certified organic" is generally left to stand on its own as self-explanatory, assisted only by general terms like "natural".

Consumer surveys have consistently identified food quality as the main reason for purchasing organic food. Higher nutritional value, no toxic residues from pesticides, and better taste are often cited, as is the positive impact of organic production on the environment. Whether organic food actually delivers on these desires and beliefs is controversial and the subject of scientifically inconclusive debate.

The debate centres on a variety of specific and supposedly demonstrable characteristics which proponents claim make organic food superior to the product of conventional farming and processing.

## Food Safety

Organic food proponents express concern over the potential negative effects of various chemical cultivation methods and genetic modification techniques used in modern conventional agriculture. The effect of pesticide residues from crop spraying, the presence of veterinary drugs in meat products, and the relatively unknown long term impacts of genetically modified varieties and breeds are all encompassed. Organic food is seen as avoiding relieving these concerns by prohibiting such practices. At present, there are no definitive scientific conclusions on any of these matters; individual studies are cited on both sides of the debate.

### *Chemical Contamination*

Organic food propcnents cite the existence of reduced levels of pesticides and herbicides as a way to reduce the long term risk of chemical consumption. The degree of risk posed by pesticide residues remains uncertain. Pesticide use in conventional food products is heavily regulated, with established, research-based maximum residue levels (MRLs) below which residues are considered safe for human consumption. Also, many pesticides are not cumulative in the body, and are regularly eliminated. Notable exceptions include heavy metals such as lead or mercury which are sometimes found in foodstuffs in countries which have lax food production standards. The U.S. and most of Europe prohibit the use of inorganic compounds containing heavy metals in any type of agriculture including conventional.

Organic farming standards do not allow the use of synthetic pesticides, but they do allow the use of certain so-called natural pesticides, such as those derived from plants. The most common organic pesticides, accepted for restricted use by most organic standards, include Bt, pyrethrum, and rotenone. Some organic pesticides, such as rotenone, have high toxicity to fish and aquatic creatures with some toxicity to mammals including humans.

A study published by the National Research Council in 1993 determined that for infants and children, the major source of exposure to pesticides is through diet. A recent study in 2006 measured the levels of organophosphorus pesticide exposure in 23 school children before and after replacing their diet with organic food. In this study it was found that levels of organophosphorus pesticide exposure dropped dramatically and immediately when the children switched to an organic diet.

One of the major differences in the production of organically produced food from traditional techniques is its reliance on so-called natural methods of providing soil nutrients. While both methods rely on nitrates to provide nitrogen to the soil, organic agriculture gets the majority of its nitrates from manure and composting sources. These sources of nitrogen have a slower transfer of nitrogen to the soil spreading out the release of nitrogen over the growing season and helping to reduce the leeching of nitrates into water sources where they can cause health problems. However, the slower transfer can also increase nitrate run-off. Manure and compost also contain the other macronutrients, phosphate and potassium, and micronutrients. While neither chemical or organic applications of nutrients to the soil provide the ideal situation, on small to medium scale operations, manure provides the benefit of reduced soil erosion and less leeching of nitrates into the soil. Regardless of the benefits, the USDA argues that the application of man-made nutrients will always be required to sustain a global agriculture system that feeds the entire world. Others, however, say that organic farming need not result in reduced production.

Organic methods of fertilization are not necessarily free from risk. However, organic animal manure, typically that of cattle, is manure from animals that eat mainly hay and other organic, primarily non-grain materials. This is seen as a way to reduce the amount of E. coli bacteria present, and the feces of organically-raised cattle have only 1% of the E. coli present in non-organic manure. It should also be noted that many organic farmers consider using manure directly as fertilizer to be an unsound practice, and instead the manure should be composted first.

### *Hormonal contamination*

Organic proponents cite evidence that some chemicals used in conventional farming, including pesticides and herbicides, mimic hormones - usually estrogen - when inside a person. They claim that this is significant even at the minute levels that the average person is exposed to. The US government states that these chemicals are safe when used correctly, but proponents claim such tests are only done on healthy adults - and that children and fetuses might be at risk to even small amounts of these chemicals.

In Australia, the Government sponsored Australian Total Diet Survey measures pesticide residues found in typical Australian diets. The 2004 survey found all estimated dietary exposures to pesticide residues were below 16% of the respective Acceptable daily intakes and therefore all exposures are well within the applicable health standards.

### *Transgenic Contamination*

The health risks surrounding genetically modified foods remain highly contentious. Other issues surrounding Genetically modified organisms (GMOs) may also concern consumers, such as the ownership of biological intellectual property by corporations, and reduction in crop varieties. Certified organic foods are not substantially genetically modified. GMO contamination of organic crops may unintentionally occur primarily through either pollen drift and/or the handling of raw organic agricultural products. There exist estimates that pollen of some crops (e.g. canola) can travel more than 5 kilometres per year.

In the United States, genetic modification is a prohibited/excluded method in organic production as defined in the National Organic Program (NOP). The USDA does not require GMO testing for all crops marketed or labelled as organic and therefore does not technically regulate GMO traces for such. The USDA does verify, through Accredited Certifying Agents, the composition and percentage of organic ingredients for all food products labelled or marketed as certified organic in the United States. The NOP does not prescribe that an Accredited Certifying Agent (ACA) revoke or suspend certification of an organic farm if accidental GMO pollen drift occurs. But all ACA's have the authority to test soils and/or crops, if necessary, for GMO's and other prohibited substances and methods. If organic certification is continued for an organic farm affected by GMO contamination (almost always through pollen drift) the farmer does have an opportunity to market most future organic crops. However, the farmers' ability to market the GMO contaminated organic crop is minimised substantially and the organic market premium is lost. In most European countries, certification rules are much stricter. Basically, any confirmed detection of transgenic plant, seed or feed can result in a loss of organic status and consequent substantial economic losses for the farmer.

### *Nutritional Value*

Some organic advocates claim that organic food is more nutritious. Increased soil quality, greater attention to quality, and selection of crop varieties for nutrition and taste instead of size, appearance, and shipping characteristics are claimed to be reasons for higher nutrient density of organic foods.

Organic advocates claim that organically grown potatoes, oranges, and leafy vegetables have more vitamin C than conventionally grown products. Phenolic compounds are also found in significantly higher concentrations in organic foods, and these may provide antioxidant protection against heart disease and cancer. Still isolated bits of research suggest that conventional agricultural practices are degrading food quality.

A study in the *Journal of the American College of Nutrition* in 2004, entitled *Changes in USDA Food Composition Data for 43 Garden Crops, 1950 to 1999*, compared vegetables analysed in 1950 and in 1999, and found noticeable decreases in six of 13 nutrients examined (the six were: protein, calcium, phosphorus, iron, riboflavin and ascorbic acid). Percentage reductions ranged from 6% for protein to 38% of riboflavin, although when evaluated on a per-food or per-nutrient level, usually no distinguishable changes were found. Reductions in calcium, phosphorus, iron and ascorbic acid were also found. The authors suggested that the differences probably reflect changes in cultivated varieties between 1950 and 1999, in which there may have been trade-offs between yield and nutrient content.

Whether organic foodstuffs have a higher nutrient content is still debatable. Studies have shown no clear, consistent results, and those that have suffer from significant experimental design flaws, according to the United Nation's Food and Agriculture Organisation (FAO). Further, the FAO found that in some crops, such as wheat, there appears to be a trade-off: in conventionally farmed wheat the levels of protein are higher, but the lower levels in organic materials are offset by gains in alpha-amylase and sugar contents.

### *Taste*

Many claim that organic food tastes better. This is primarily referred to regarding fresh food. It is possible that organic food tastes better simply because it is fresher. Because organic farms tend to be smaller, they often sell their products closer to the point of harvest. Thus, organic fruits and vegetables taste more "farm fresh" than comparable conventional produce. However, organic foods might also have more flavour because organic farmers often breed with taste instead of marketability as the primary factor.

Conventional tomatoes, for example, are often bred to be perfectly red and round, to match the ideal appearance of a tomato. They are also bred to resist damage in transport and storage, for a longer shelf-life. This means that taste is an attribute that has a lower priority. In addition to crop diversity and selection practices, organic farming emphasizes soil nutrition, which can positively influence the taste of the food. Tests by the United Nations FAO demonstrated that some apples, specifically the "Golden Delicious" variety, have higher flavonoid counts when grown organically. This suggests that they do have more flavour. Some foods, such as bananas, are picked when unripe, then artificially induced to ripen using a chemical (such as propylene or ethylene) while in transit, possibly producing a different taste. The issue of ethylene use in organic food production is contentious; opponents claiming that its use only benefits large companies, and opens the door to weaker organic standards.

### *Environmental Impact*

Every food purchase supports the system that delivers it, and if large-scale chemical production methods are damaging to the environment, then purchasing these foods supports this damage. A main goal of organic farming is minimising impact to the environment.

### *Sustainability*

Proponents of organic farming say that "conventional" farming is unsustainable, because it relies on artificial inputs (synthetic fertilizers and other chemicals, machinery, etc.) that ultimately requires energy in the form of fossil fuels, and because the land is degraded through soil erosion, salinization, and other processes that eventually render the soil infertile.Many claim that without cheap fossil fuels and government subsidies, conventional agriculture would not be possible, and that despite technological advancements, there will eventually be an agricultural crisis as a result of depleted soil. The cultivation of monocultures, many acres planted with the same crop year after year, increases susceptibility to pests and diseases and depletes the soil, while eliminating most native flora and fauna. In contrast, organic farming often utilises intercropping, crop rotation, fallow periods, and integrated pest management to promote biodiversity and preserve the health of the soil while minimising the risk of diseases. The main goal of organic farming is sustainability, so organic farms seek to minimise dependence on outside resources and be self-sufficient.

### *Pollution*

Modern agricultural practices often result in large amounts of nitrogen runoff from the heavy use of fertilizer, which pollutes watersheds. In addition to posing a threat to human health and disrupting aquatic ecosystems, this sometimes results in algal blooms which deplete the water of oxygen resulting in fishkills. Pesticide runoff also causes many problems.

### *Transportation*

Claims that eating organic food is better for the environment are, however, frustrated by the fact that most of the organic food sold today travels the same great distances as conventional food. A UK study published in 2005 in the *Journal of Food Policy* found that maximum environmental benefit would result from purchasing food produced within a 12-mile radius. Therefore, buying local food that is not organic could be environmentally "better" than buying organic food that has travelled hundreds or thousands of miles. Many small organic farms, however, sell much or all of their produce locally.

### *Energy Efficiency*

Studies have shown organic farms to be more energy efficient than their conventional counterparts. One of these studies was done with apple farms in the state of Washington. In that study, the organic farms were found to be at least 7% more energy efficient. Although some critics of organic farms cite evidence that organic farms produce less yield than conventional farms, this study also found a much more substantial decrease in resources used. One prominent 21-year Swiss study found average 20% lower organic yields over conventional methods. However, that came with consumption of 50% less fertilizer, and 97% less pesticide.

In comparing yields, a US survey published in 2001 analysed 150 growing seasons of data on various crops and concluded that organic yields were 95-100% of conventional yields. Because organic farms don't use toxic pesticides and herbicides, there is more biodiversity in the soil. Besides higher soil quality - more life in the soil allows for higher water retention. This helps increase yields for organic farms in drought years where there is less rain. During drought years, organic farms have been found to have yields 20-40% higher than conventional farms.

### *Large Scale Organic Farms*

Many advocates of organic farming view large scale, corporate owned "organic farms" as being against the spirit of organic farming, since they tend to use unsustainable practices similar to conventional farms.

## Organic Food

*Organic foods* are foods that are produced using methods that do not involve modern synthetic inputs such as synthetic pesticides and chemical fertilizers. Organic foods are not processed using irradiation, industrial solvents, or chemical food additives. The weight of the available scientific evidence has not shown a consistent and significant difference between organic and more conventionally grown food in terms of safety, nutritional value, or taste. For the vast majority of human history, agriculture can be described as organic; only during the 20th century was a large supply of new synthetic chemicals introduced to the food supply. The organic farming movement arose in the 1940s in response to the industrialization of agriculture known as the Green Revolution.

Organic food production is a heavily regulated industry, distinct from private gardening. Currently, the European Union, the United States, Canada, Japan and many other countries require producers to obtain special certification in order to market food as organic within their borders. In the context of these regulations, organic food is food produced in a way that complies with organic standards set by national governments and international organisations.

In the United States, organic production is a system that is managed in accordance with the Organic Foods Production Act (OFPA) of 1990 and regulations in Title 7, Part 205 of the Code of Federal Regulations to respond to site-specific conditions by integrating cultural, biological, and mechanical practices that foster cycling of resources, promote ecological balance, and conserve biodiversity. If livestock are involved, the livestock must be reared with regular access to pasture and without the routine use of antibiotics or growth hormones. In the United States, a food can be labelled as "organic" if it contains a minimum of 95% organic ingredients. In most countries, organic produce also do not contain genetically modified organisms. It has been suggested that the application of nanotechnology to food and agriculture is a further technology that needs to be excluded from certified organic food. The Soil Association (UK) has been the first organic certifier to implement a nano-exclusion.

### Meaning and Origin of the Term

In 1939, Lord Northbourne coined the term *organic farming* in his book *Look to the Land* (1940), out of his conception of "the farm as organism," to describe a holistic, ecologically balanced approach to farming—in contrast to what he called *chemical farming*, which relied on "imported fertility" and "cannot be self-sufficient nor an organic whole." This is different from the scientific use of the term "organic," to refer to an a class of molecules that contain carbon, especially those involved in the chemistry of life. This class of molecules includes everything likely to be considered edible, and include most pesticides and toxins too, therefore the term "organic" and, especially, the term "inorganic" (sometimes wrongly used as a contrast by the popular press) are both technically inaccurate and completely inappropriate when applied to farming, the production of food, and to foodstuffs themselves.

### *Identifying Organic Food*

Processed organic food usually contains only organic ingredients. If non-organic ingredients are present, at least a certain percentage of the food's total plant and animal ingredients must be organic (95% in the United States, Canada, and Australia) and any non-organically produced ingredients are subject to various agricultural requirements. Foods claiming to be organic must be free of artificial food additives, and are often processed with fewer artificial methods, materials and conditions, such as chemical ripening, food irradiation, and genetically modified ingredients. Pesticides are allowed as long as they are not synthetic.

Early consumers interested in organic food would look for non-chemically treated, non-use of unapproved pesticides, fresh or minimally processed food. They mostly had to buy directly from growers: "Know your farmer, know your food" was the motto. Personal definitions of what constituted "organic" were developed through first-hand experience: by talking to farmers, seeing farm conditions, and farming activities. Small farms grew vegetables (and raised livestock) using organic farming practices, with or without certification, and the individual consumer monitored.

As demand for organic foods continued to increase, high volume sales through mass outlets such as supermarkets rapidly replaced the direct farmer connection. Today there is no limit to organic farm sizes and many large corporate farms currently have an organic division. However, for supermarket consumers, food production is not easily observable, and product labelling, like "certified organic", is relied on. Government regulations and third-party inspectors are looked to for assurance. The USDA carries out routine inspections of farms that produce USDA Organic labelled foods. On April 20, 2010, the Department of Agriculture said that it would begin enforcing rules requiring the spot testing of organically grown foods for traces of pesticides, after an auditor exposed major gaps in federal oversight of the organic food industry.

### *Environmental Impact*

Several surveys and studies have attempted to examine and compare conventional and organic systems of farming. The general consensus across these surveys is that organic farming is less damaging for the following reasons:

- Organic farms do not consume or release synthetic pesticides into the environment—some of which have the potential to harm soil, water and local terrestrial and aquatic wildlife although it must be noted that organic pesticides still have the potential to be as damaging to the environment as synthetic pesticides.
- Organic farms are better than conventional farms at sustaining diverse ecosystems, *i.e.*, populations of plants and insects, as well as animals.
- When calculated per unit area, organic farms use less energy and produce less waste, *e.g.*, waste such as packaging materials for chemicals.

A 2003 investigation by the Department for Environment Food and Rural Affairs in the UK found, similar to other reports, that organic farming "can produce positive environmental benefits", but that some of the benefits were decreased or lost when comparisons are made on "the basis of unit production rather than area".

The situation was shown dramatically in a comparison of a modern dairy farm in Wisconsin with one in New Zealand in which the animals grazed extensively. Using total farm emissions per kg milk produced as a parameter, the researchers showed that production of methane from belching was higher in the New Zealand farm, while carbon dioxide production was higher in the Wisconsin farm. Output of nitrous oxide, a gas with an estimated global warming potential 310 times that of carbon dioxide was also higher in the New Zealand farm. Methane from manure handling was similar in the two types of farm.

The explanation for the finding relates to the different diets used on these farms, being based more completely on forage (and hence more fibrous) in New Zealand and containing less concentrate than in Wisconsin. Fibrous diets promote a higher proportion of acetate in the gut of ruminant animals, resulting in a higher production of methane that has to be released by belching. When cattle are given a diet containing some concentrates (such as corn and soybean meal) in addition to grass and silage, the pattern of ruminal fermentation alters from acetate to mainly propionate. As a result methane production is reduced. Capper et al. compared the environmental impact of US dairy production in 1944 and 2007. They calculated that the carbon "footprint" per billion kg of milk produced in 2007 was 37 percent that of equivalent milk production in 1944.

The environmental impact of pesticides as well as the impact to the health of farm workers is reasons given for purchasing organic food. However, critics of organic farming methods believe that the increased land needed to farm organic food could potentially destroy the rainforests and wipe out many ecosystems.

### *Yield*

One study found a 20% smaller yield from organic farms using 50% less fertilizer and 97% less pesticide. Studies comparing yields have had mixed results. Supporters claim that organically managed soil has a higher quality and higher water retention. This may help increase yields for organic farms in drought years.

One study from the Danish Environmental Protection Agency found that, area-for-area, organic farms of potatoes, sugar beet and seed grass produce as little as half the output of conventional farming. Michael Pollan, author of *The Omnivore's Dilemma*, responds to this by pointing out that the average yield of world agriculture is substantially lower than modern sustainable farming yields. Bringing average world yields up to modern organic levels could increase the world's food supply by 50%.

A 2007 study compiling research from 293 different comparisons into a single study to assess the overall efficiency of the two agricultural systems has concluded that...organic methods could produce enough food on a global per capita basis to sustain the current human population, and potentially an even larger population, without increasing the agricultural land base.

The researchers also found that while in developed countries, organic systems on average produce 92% of the yield produced by conventional agriculture, organic systems produce 80% more than conventional farms in developing countries, because the materials needed for organic farming are more accessible than synthetic farming materials to farmers in some poor countries. On the other hand, communities that lack sufficient manure to replenish soils would struggle with organic farming, and the soil would degrade rapidly.

### *Energy Efficiency*

A study of the sustainability of apple production systems showed that in comparing a conventional farming system to an organic method of farming, the organic system in this case is more energy efficient. A more comprehensive study compared efficiency of agriculture for products such as grain, roughage crops, and animal husbandry. While the study did not investigate specific additional requirements of arable land or numbers of farm labourers to produce total yields for organic farming vs. conventional farming, leaving open the question of overall capacity of organic farming to meet current and future agricultural needs, it concluded that organic farming had a higher yield per unit of energy over multiple crops and for livestock. However, conventional farming had higher total yield. Conversely, another study noted that organic wheat and corn production was more energy efficient than conventional methods while organic apple and potato production was less energy efficient than conventional methods. A long-term study, spanning two decades, noted that crop yields were 20% lower in organic systems while fertilizer plus energy input was 34% to 53% lower. However, pesticide input was reduced by 97% in organic farm systems.

### *Health and Safety*

The Research Institute of Organic Agriculture in Switzerland in 2009 concluded that "there is broad agreement on two points: there is no proof that organic food is more nutritious or safer, and most studies that have compared the taste and organoleptic quality of organic and conventional foods report no consistent or significant differences between organic and conventional produce."

### *Consumer Safety*

Reviews of the available body of scientific literature have not found that organic food is any safer or healthier than conventional foods. There continues to be widespread public belief in arguments that organic food is significantly safer for consumption than food grown conventionally, based mainly on anecdotal evidence and testimonials rather than scientific evidence; this belief has fuelled increased demand for organic food despite higher prices. Firm conclusions about the relative safety of organic foods have been hampered by the difficulty in proper study design and relatively small number of studies directly comparing organic food to conventional food.

Claims of improved safety of organic food have largely focused on pesticide residues. While studies have shown organically grown fruits and vegetables have significantly lower pesticide residue levels, the significance of this finding on actual health risk reduction is debatable as both conventional foods and organic foods generally have pesticide levels well below government established guidelines for what is considered safe. This view has been echoed by the U.S. Department of Agriculture and the UK Food Standards Agency. Claims of increased risk related to pesticide residue and rates of infertility or lower sperm counts have not been supported by the evidence in the medical literature. Reviews have noted that the risks from microbiological sources or natural toxins are likely to be much more significant than short term or chronic risks from pesticide residues.

Some focus has been placed on the amount of nitrogen content in certain vegetables, especially green leafy vegetables and tubers when grown organically as compared to conventionally. While these vegetables, when grown organically, have been found to have lower nitrogen content, there is no consensus as to whether consumption of lower levels of nitrogen translates to improved health risk. When evaluating environmental toxins such as heavy metals, the USDA noted that organically raised chicken may have lower arsenic levels, while literature reviews found no significant evidence that levels of arsenic, cadmium or other heavy metals differed significantly between organic and conventional food products.

In looking at possible increased risk to safety from organic food consumption, reviews have found that although there may be increased risk from microbiological contamination due to increased manure use as fertilizer from organisms like *E. coli* O157:H7 during organic produce production, there is little evidence of actual incidence of outbreaks which can be positively blamed on organic food production. One outbreak of E. coli in Germany was blamed on organic farming of bean sprouts. Other possible sources of increased safety risk from organic food consumption like use of biological pesticides or the theoretical risk from mycotoxins from fungi grown on products due to the lack of effective organic compliant fungicides have likewise not been confirmed by rigorous studies in the scientific literature. The American Cancer Society (ACS) has noted that interest in organic food is derived from perceived risk of cancer from additives not found in organic foods. The ACS has stated their official position that "whether organic foods carry a lower risk of cancer because they are less likely to be contaminated by compounds that might cause cancer is largely unknown" and "vegetables, fruits, and whole grains should form the central part of a person's diet, regardless of whether they are grown conventionally or organically".

### *Nutritional Value and Taste*

According to the UK's Food Standards Agency, "Consumers may choose to buy organic fruit, vegetables and meat because they believe them to be more nutritious than other food. However, the balance of current scientific evidence does not support this view." A 12-month systematic review commissioned by the FSA in 2009 and conducted at the London School of Hygiene & Tropical Medicine based on 50 years' worth of collected evidence concluded that "there is no good evidence that consumption of organic food is beneficial to health in relation to nutrient content." Other studies have found no proof that organic food offers greater nutritional values, more consumer safety or any distinguishable difference in taste. A review of nutrition claims showed that organic food proponents are unreliable information sources which harm consumers, and that consumers are wasting their money if they buy organic food believing that it contains better nutrients. Minor differences in ascorbic acid, protein concentration and several micronutrients have been identified between organic and conventional foods, but it doesn't appear that these have any impact on human health.

As with vegetable produce, there is evidence that some organic fruit is drier than conventionally grown fruit. Unless this factor is taken into account a higher content of a nutrient might be explained by a higher dry-matter (lower moisture) content. A slightly drier fruit may also have a more intense flavour due to the higher concentration of nutrients, and as a result may be preferred by the consumer. There is evidence that some organically grown fruits has a higher resistance to deterioration and better keeping quality, attributed to a lower moisture content.

Although it is commonly claimed that organically grown food tastes better than conventionally grown food, reviews of the literature that looked at the sensory qualities of the two have not found convincing evidence that there are any significant differences.

Psychological effects such as the "halo" effect which are related to the choice and consumption of organic food and which may be akin to religious experiences in some people are in addition important motivating factors in the purchase of organic food. An example of the halo effect was demonstrated by Schuldt and Schwarz. Their results showed that university students inferred that organic cookies were lower in calories and could be eaten more often than conventional cookies. This effect was observed even when the nutrition label conveyed an identical calorie content. The effect was more pronounced among participants who were strong supporters of organic production and had strong feelings about environmental issues. The perception that organic food is low-calorie food or health food appears to be quite common.

Demand for organic foods is primarily concern for personal health and concern for the environment. Organic products typically cost 10 to 40% more than similar conventionally produced products.

According to the USDA, Americans, on average, spent $1,347 on groceries in 2004; thus switching entirely to organics would raise their cost of groceries by about $135 to $539 per year ($11 to $45 per month) assuming that prices remained stable with increased demand. Processed organic foods vary in price when compared to their conventional counterparts.

While organic food accounts for 1–2% of total food sales worldwide, the organic food market is growing rapidly, far ahead of the rest of the food industry, in both developed and developing nations.

- World organic food sales jumped from US $23 billion in 2002 to $52 billion in 2008.
- The world organic market has been growing by 20% a year since the early 1990s, with future growth estimates ranging from 10%–50% annually depending on the country.
- Organic food is the fastest growing sector of the American food marketplace.
- Organic food sales have grown by 17 to 20 percent a year for the past few years while sales of conventional food have grown at only about 2 to 3 percent a year.
- In 2003 organic products were available in nearly 20,000 natural food stores and 73% of conventional grocery stores.
- Organic products accounted for 3.7% of total food and beverage sales, and 11.4% of all fruit and vegetable sales in the year 2009.
- Two thirds of organic milk and cream and half of organic cheese and yogurt are sold through conventional supermarkets.
- As of 2012, most independent organic food processors in the USA had been acquired by multinational firms.
- Organic food sales surpassed $1 billion in 2006, accounting for 0.9% of food sales in Canada.
- Organic food sales by grocery stores were 28% higher in 2006 than in 2005.
- British Columbians account for 13% of the Canadian population, but purchased 26% of the organic food sold in Canada in 2006.

In the European Union (EU25) 3.9% of the total utilised agricultural area was used for organic production in 2005. The countries with the highest proportion of organic land were Austria (11%) and Italy (8.4), followed by the Czech Republic and Greece (both 7.2%). The lowest figures were shown for Malta (0.1%), Poland (0.6%) and Ireland (0.8%).

In 2009, the proportion of organic land in the EU grew to 4.7%. The countries with highest share of agricultural land were Liechtenstein (26.9%), Austria (18.5%) and Sweden (12.6%). 16% of all farmers produced organically in 2010. By the same year the proportion of organic land increased to 20%. In 2011, 7.4% of all food products sold in Austrian supermarkets (including discount stores) were organic. In 2007, 8,000 different organic products were available.

Since 2000, the use of some organic food is compulsory in Italian schools and hospitals. A 2002 law of the Emilia Romagna region implemented in 2005, explicitly requires that the food in nursery and primary schools (from 3 months to 10 years) must be 100% organic, and the food in meals at schools, universities and hospitals must be at least 35% organic.

In 2005 168,000 ha of land were under organic management. 7 percent of Polish consumers buy food that was produced according to the EU-Eco-regulation. The value of the organic market is estimated at 50 million Euros (2006). 100,000 ha of land were under organic management in 2010, representing 1% of the total utilised agricultural area. 70%-80% of the local organic production, amounting to 100 million Euros in 2010, is exported. The organic products market grew to 50 million Euros in 2010.

Organic food sales increased from just over £100 million in 1993/94 to £1.21 billion in 2004 (an 11% increase on 2003). In 2010, the UK sales of organic products fell 5.9% to £1.73 billion. 86% of households buy organic products, the most popular categories being dairies (30.5% of sales) and fresh fruits and vegetables (23.2% of sales). 4.2% of UK farmland is organically managed.

After the collapse of the Soviet Union in 1991, agricultural inputs that had previously been purchased from Eastern bloc countries were no longer available in Cuba, and many Cuban farms converted to organic methods out of necessity. Consequently, organic agriculture is a mainstream practice in Cuba, while it remains an alternative practice in most other countries. Although some products called organic in Cuba would not satisfy certification requirements in other countries (crops may be genetically modified, for example), Cuba exports organic citrus and citrus juices to EU markets that meet EU organic standards. Cuba's forced conversion to organic methods may position the country to be a global supplier of organic products.

## Natural Foods

"Natural foods" and "all natural foods" are widely used terms in food labelling and marketing with a variety of definitions, most of which are vague. The term is assumed to imply foods that are minimally processed and do not contain manufactured ingredients, but the lack of standards in most jurisdictions means that the term assures nothing. The term "organic" has similar implications and has an established legal definition in many countries and an international standard. In some places, the term "natural" is defined and enforced. In others, such as the United States, it has no meaning.

### *Formal Definitions, or Lack Thereof*

"Natural foods" are often assumed to be foods that are minimally processed and do not contain any hormones, antibiotics, sweeteners, food colours, or flavourings that were not originally in the food. The terms are often misused on labels and in advertisements. The international Food and Agriculture Organisation's *Codex Alimentarius* does not recognise the term "natural" but does have a standard for organic foods.

Fundamentally, almost all foodstuffs are derived from the natural products of plants and animals; therefore, any definition of "natural food" results in an arbitrary exclusion or inclusion of food ingredients; likewise, since almost all foods are processed in some way, either mechanically, chemically, or by temperature, it is difficult to define which types of food processing are "natural".

In the United Kingdom, the Food Standards Agency has published criteria for the use of several terms in food labelling. The guidance, in general, restricts the use of natural to foods that have "ingredients produced by nature, not the work of man or interfered with by man." Natural flavourings are explicitly defined by separate laws.

There are different standards for various types of food, such as dairy products. It also gives standards for some food processing techniques, such as fermentation or pasteurization.

The standard explicitly rules out "foods derived from novel processes, GM or cloning." The Canadian Food Inspection Agency restricts the use of "natural" to foods that have not been significantly altered by processing and gives examples of processes that do or do not significantly alter food. This includes two specific additional requirements:

- A natural food or ingredient of a food is not expected to contain, or to ever have contained, an added vitamin, mineral nutrient, artificial flavouring agent or food additive.
- A natural food or ingredient of a food does not have any constituent or fraction thereof removed or significantly changed, except the removal of water.

In Israel, natural ingredients are defined as part of the Labelling of Prepacked Food Standard (Israeli Standard SI 1145, which is legally binding).

The standard offers a list of 33 processes which are allowed in natural ingredients, all of which are physical treatments and not chemical modifications. These include blending, cleaning, extrusion, freezing, drying, etc. A specific ingredient can be called "natural" if it didn't go through any processing except for the listed ones. The whole food can be called "natural" if the food is not a blend of foods (even if they are all natural), has no added ingredients, and underwent only the specified processes.

In the United States, neither the Food and Drug Administration (FDA) nor the U.S. Department of Agriculture (USDA) has rules for "natural." The FDA explicitly discourages the food industry from using the term. The Food, Drug, and Cosmetic Act prohibits labelling that is false or misleading, but does not give any specifics. The USDA's Agricultural Marketing Service has a standard for organic food.

Because there is no legal meaning for natural foods, food manufacturers can include ingredients that may not be considered natural by some consumers. The poultry industry has been criticized by the Centre for Science in the Public Interest for labelling chicken meat "all natural" after it has been injected with saline solution up to 25% of its weight, but there is no legal recourse to prevent this labelling. Although there is no legal U.S. definition for natural foods, there are numerous unofficial or informal definitions, none of which is applied uniformly to foods labelled "natural".

## Organic Beans

Organic Beans are produced and processed without the use of synthetic fertilizers and pesticides. In 2008, over 2,600,000 acres (11,000 $km^2$) of cropland were certified organic in the United States. Dry beans, snap beans, and soybeans were grown on 16,000 acres (65 $km^2$), 5,200 acres (21 $km^2$), and 98,000 acres (400 $km^2$), respectively. There are three major types of organic beans: dry beans, snap beans (also known as greenbeans), and soybeans.

The mature seed of dry beans (*Phaseolus vulgaris*) serve as a protein source in a variety of foods worldwide. Dry beans and snap beans are the same species, although dry beans are distinguished from snap beans (greenbeans) which are consumed as immature pods. Dry and snap beans also differ from soybeans (*Glycine max*) in which the seeds are consumed in a variety of processed forms such as tofu, soybean meal, oil, and fermented forms such as miso. Dry beans are divided into classes with a range of seed sizes, colours, and shapes. Examples include black beans, pintos, navy beans, small red beans, and kidney beans. Soybeans are usually divided into two groups feed-grade and food-grade, with the food-grade including soybeans for processing, "tofu-beans", and edamame. The latter of which is eaten as immature beans.

### *Organic Certification*

To sell beans as organic in the United States, producers must meet the requirements of the United States Department of Agriculture (USDA) National Organic Program (NOP) under the Organic Food Production Act of 1990. A third party certifying agency, not part of the USDA, verifies that the producer has meet the minimum requirements and may themselves have additional requirements.

Prior to marketing crops as "certified organic", fields must be managed during a 3-year transition phase using organic practices (that is, no synthetic pesticides or fertilizers). Detailed record keeping is essential for organic producers and these details are checked annually by the certifying agency who may perform a site inspection. Certification information for other nations can be found on the Organic Certification wikipedia page.

### *Production Information*

Michigan is currently the top organic dry bean producing state in the U.S., accounting for 37% of the area and 47% of the sales. The majority of the area is located in the "Thumb" region of Michigan, with black beans being the most common class produced organically. Colorado and North Dakota are the second and third top producers, respectively, of organic dry beans in the United States.Washington is the top producer of organic snap beans, followed by Michigan and California. In 2008, the total sales of organic snap beans totalled over 1.4 million dollars. The value of organic soybeans in 2008 totalled more than $50 million dollars in the U.S., with Minnesota, Iowa, and Michigan accounting for 46% of the total production.

### *Organic Bean Markets*

Organically grown beans typically demand a price premium and as a consequence the marketing details differ from the conventional market. Organic bean producers often set up contracts with buyers prior to planting the crop . Buyers may also have special needs to consider when it comes to cleaning and processing beans in certified facilities, thereby making proximity to these certified facilities important. Organic dry beans from the United States are marketed both nationally and abroad. In 2010, one of the markets for Michigan and Washington produced dry beans was the restaurant chain Chipotle Mexican Grill. Another market is Eden Foods Inc. Organic snap beans are marketed to a variety of companies, such as Gerber and Earth's Best, makers of organic baby food. Organic soybeans have perhaps the most market potential of the three beans as they can be used as feed for organic animal production and processed into many different products. Eden Foods Inc. is one example of a market for organically produced soybeans.

### *Pest Management*

Pest management in organic dry beans must be done without the use of synthetic herbicides, fungicides, insecticides, and other pesticides. For this reason, cultural, mechanical, and biological methods are the primary means of keeping pests under control.

### *Major Weed Pests*

Controlling weeds is the primary cost to agricultural production systems, including organic beans. Weeds can reduce the emergence, growth, and yield of beans as they compete for light, water, and nutrient resources. Weeds are grouped in many ways, such as by life cycle. In organic beans the most problematic weeds are those with summer annual and perennial life cycles.

### *Summer Annuals*

Summer annuals are weeds that germinate in the spring and/or summer, produce seed in the late-summer to fall, and then die. Their lifecycle is complete in one year.

- Common lambsquarters- broadleaf weed that emerges over an extended period of time, grows rapidly, produces large quantities of seed that can remain dormant in the soil for decades
- Common ragweed- broadleaf weed that emerges over an extended period of time, grows rapidly, and is tolerant to control by heat treatments (like a propane flamer)
- Pigweed (*Amaranthus*)- broadleaf weed that grows rapidly and is a prolific producer of seed
- Foxtail (*Setaria*) species- grass weed that grows in clumps and emerges after common lambsquarters and common ragweed
- Crabgrass (Digitaria) species- grass weed with a prostrate growth habit.

### *Perennials*

Perennial weeds can live for multiple growing seasons, usually due to hardy root stocks.

- Canada thistle- broadleaf weed with roots that can grow to a depth 15 feet (4.6 m) in the soil profile. Tillage often promotes the spread of Canada thistle by dispersing segments of the root.
- Perennial sowthistle- broadleaf weed with yellow flowers similar to dandelion that can reproduce by seed or rhizome

- Quackgrass- grass weed that reproduces primarily via rhizome. Seeds are short lived in the soil (2–4 years)

Example of weed management: Several types of tractor-pulled implements such as a rotary hoe and/or one of various types of cultivator/cultivators can be used to mechanically remove weeds from organic fields.

### *Major Insect Pests*

Insects are another contributor to yield loss in organic beans. There are several examples of insects that affect organic bean systems including:

- Bean and potato leafhoppers- sucking insect that causes bean leaves to yellow at the tips and on the edges and can cause stunting
- Western bean cutworm- larvae of this insect feed on the pods of beans reducing yield and quality
- Soybean aphid- sap sucking insect that can form large colonies on soybean. Soybean aphids can also transmit viruses from plant to plant while feeding.
- Seedcorn maggot- feed on dry and snap bean seed and seedlings. The risk of feeding is increased when temperatures are cool and the soil is wet.

Example of insect management: *Bacillus thuringiensis*(Bt) is a bacterium that when ingested by susceptible insects is lethal. This product is referred to as a microbial insecticide of which there are four subspecies and over 100 commercial products available for use in organic systems.

### *Major Pathogens*

Pathogens can take the form of fungi, bacteria, viruses, and nematodes. Each can act as a disruptor to plant growth and development and can have negative effects on organic bean yields. Examples of bean pathogens include:

- White mold- a fungus with a wide host range. A symptom of later stages of infection is white cottony hyphal growth on bean stems, leaves, and pods.
- Anthracnose- refers to disease caused by many different fungi. Anthracnose spores can be seedborne or overwinter on leaf litter.

- Soybean cyst nematode (soybean specific)- can cause up to 50% yield loss in soybeans due to root feeding.
- Bean common and soybean mosaic viruses- virus that occurs on bean seed and can have negative impacts on yield and bean quality. Both viruses can be trasmitted from plant to plant by aphids.

Example of disease management: The onset of certain diseases, such as white mold, can be reduced by planting in rows wide enough to allow adequate air movement which can reduce humidity and decrease drying time after precipitation events. If plant material is allowed to dry quickly it will not have the 9 to 48 hours of continuous moisture on the leaf surface required for the white mold infection to occur.

### *Organically Approved Pesticides*

Some naturally derived chemical products are permitted for use under organic production. All materials have to be reviewed and approved by the Organic Materials Review Institute (OMRI). Examples of organically allowable pesticides that can be used to produce organic beans include:

- Neem oil- Used as an insecticide
- Acetic acid- Used as a herbicide
- Pyrethrum- Broad spectrum insecticide derived from *Chrysanthemum*
- *Bacillus subtilis*- Used as a fungicide
- *Bacillus thuringiensis*- Used as an insecticide.

### *Fertility*

Because synthetic fertilizers are not permitted in organic production systems, fertility must come from sources such as:

- Green manures (cover crops)
- Composts and compost teas
- Livestock manures
- Rock minerals- such as limestone, gypsum, and rock phosphate

Beans are just one group of organically produced agricultural products. View the section below for other organic agriculture related Wikipedia entries.

## Organic Clothing

Organic clothing is clothing made from materials raised or grown in compliance with organic agricultural standards. Organic clothing uses cotton, jute, silk, ramie, or wool. Retailers charge more for organic clothing because such the materials used are free from chemicals or genetically modified seeds. Alternative terms are organic fashion, organic cloths.

### *Benefits*

Authentic organic fabrics and clothing can help the environment in a number of ways, such as:

- Manufacture of chemicals is not required
- Chemical residues are not entered accidentally into the environment
- Humans and animals are not exposed to chemicals
- When the fabric is finished with chemicals are not returned to the earth in landfill, or enter into recycling process.

Cotton covers 2.5% of the world's cultivated land yet uses 16% of the world's insecticides, more than any other single major crop. It can take almost a 1/3 pound of synthetic fertilizers to grow one pound of raw cotton in the US, and it takes just under one pound of raw cotton to make one t-shirt.

Many highstreet retailers market organic clothing ranges that contain chemicals from the dyeing to bleaching process, which is inconsistent with the idea of Organic Clothing. Many companies sell clothing made from bamboo, which is commonly labelled as "organic", however this is a false statement. Bamboo fabric is typically chemically manufactured by "cooking" the bamboo leaves and woody shoots in strong chemical solvents such as sodium hydroxide and carbon disulfide, in a process also known as hydrolysis alkalization combined with multi-phase bleaching.

Both sodium hydroxide and carbon disulfide have been linked to serious health problems. This finished material is similar to rayon and modal, which are more accurate terms of describing bamboo fabrics. Criticism also concerns the high cost of the products.

# 7

# Acquisition and Adoption of Organic Farming Practices

Beginning in the 1990s, new considerations have been added to agricultural policies worldwide. An increasing number of countries, including those of the European Union (EU), have begun to: recognise the need for introducing the principle of sustainability in their policies concerning the use of agricultural and natural resources, and liberalize their agricultural sectors by reducing support policies and dismantling agricultural trade impediments. Initiated at the Uruguay Round on Trade, and strengthened by the founding of the World Trade Organisation (WTO), this course is expected to continue in the upcoming Negotiation Rounds on Trade. To conform with these developments, the EU has already taken steps in adjusting its Common Agricultural Policy (CAP).

One of the means utilised by the EU to keep up with these developments is the introduction of standards (such as quality and environmental standards) in farming. Practically, this has been pursued by institutionalizing, via EU regulations, techniques for producing differentiated versions of agricultural products such as Protected Designations of Origin (PDOs), Protected Geographical Indication (PGI), and organically produced commodities.

Among these, organic farming, institutionalized via EU Regulation 2092/1991 as amended by Regulation 1804/1999, represents a promising alternative for the future of European agriculture for at least three reasons: first, it is consistent with the notion of sustainable development set forth in the 1992

CAP Reform; second, it provides a solution to falling farm income observed during the last decade because of the existing production surpluses; and third, organically produced commodities constitute an appealing option for weary EU consumers in the wake of alarming food-safety events such as "mad cow" disease and dioxin-poisoned food.

However, despite the widespread interest in organic agriculture, it still represents only a small portion of the total utilised agricultural area in most European countries. This modest participation in organic agriculture is not surprising. Organic farming is a risky business for a newcomer, as it introduces a number of uncertainties. Among other things, farmers are uncertain both about how much output they will be able to produce for given inputs and about the prices they will be able to secure for the inputs and the output. Such uncertainties may lead to ill-informed production decisions, which are not only detrimental to the well-being of the farmer but also affect the future course of organic farming in general.

In order to cope with the problem of low adoption rates, several European countries have promoted this mode of farming via mainly subsidy-driven policies which are summarised in EU Regulation 1257/1999. Specifically, direct subsidy schemes were introduced requiring conversion of at least a portion of a farm's land and continued organic production.

In an analysis of these policy schemes, Lampkin and Padel (1994) found that conversion subsidies expanded organic farming significantly throughout Europe, at least in the early years. Indeed, financial incentives such as direct subsidies (whereby the central government essentially "shares" the risk of adoption) are common and effective means of overcoming farmers' adverse perceptions. These types of incentives are costly, however, especially if adoption depends primarily on perceptions about future yields. In addition, direct financial support schemes cannot ensure the economic viability of organic farm operations in the long run. Moreover, they are in sharp contrast with the recently initiated processes of agricultural market liberalization and reduction of price support and production grants.

A promising and equally effective way to promote technological adoption in the farming sector is the improvement of farmers' allocative ability through the provision of informational incentives that revise their perceptions about the profit-effectiveness of new farming technologies. Although fixed initial costs are incurred, informational incentives may be less costly than financial incentives in the long run as information spreads throughout the rural communities. While both information and subsidy policies speed up adoption and diffusion of new technologies, Stoneman and David (1986) have shown that subsidy policies may yield welfare losses in the form of income transfers from other sectors of the economy.

Moreover, in a recent study analysing EU policies related to organic farming, Lohr and Salomonsson (2000) found [in contrast with Lampkin and Padel (1994); Pietola and Oude Lansink (2001); and Musshoff and Odening (2005)] that market services and information sources rather than subsidies are more effective in encouraging organic adoption throughout the EU. Although the relevant EU Regulations include various measures to provide farmers with the necessary information required to improve their respective expertise on organic technologies (e.g., extension provision), subsidy-driven policies have remained the primary incentive for organic conversion throughout the EU. It follows, therefore, that farmers' attitudes toward actively seeking (or not) information about their professional activities are of major importance for the organic adoption decision.

In light of the above, the objective of this chapter is to offer an empirical analysis of how information from various sources impacts farmers' allocative ability and thus their decisions to adopt organic farming practices, and what characteristics of farmers and farm businesses may have an effect on their decisions to acquire more information and begin organic production. To achieve this goal, we extend Wozniak's (1993) estimation procedure from a simple correlation to a structural model using a multivariate probit estimator. In particular, we specify the farmer's organic adoption and farming information-gathering decisions as a recursive simultaneous trivariate ordered probit model which we apply to a cross-sectional data set of Cretan farms.

The next section details the theoretical framework for jointly analysing the technological adoption and information acquisition processes. The resulting econometric specification is then presented, followed by a description of the data. Next, the estimation results are discussed, in combination with some policy recommendations implied by our findings. Summary remarks are offered in the final section.

The farmer's decision to adopt technological innovations is an issue extensively studied since the publication of Griliches' (1957) pioneering work on the adoption of hybrid corn in the United States. The major body of the existing economic research on technology adoption has been concerned with the question of what determines the decision of a farmer to adopt or reject an innovation. However, there is a relative dearth of empirical research in addressing the link between the farmer's decision to adopt innovations and his or her decision to gather information not only on new technologies available, but also on farming practices in general. Accordingly, we assume a farmer's decision to "turn organic" may be influenced by his or her general information acquisition process, in the sense that this process might induce a shift in the probability of adopting the new technology. According to human capital theory, innovative ability is closely related to education level, experience, and information accumulation-i.e., those characteristics associated with the resource allocation skills of farm operators. Information gathering, regardless of whether or not this refers to the innovation itself, is expected to enhance resource allocation skills and to increase the efficiency of adoption decisions.

Farmers with a high level of resource allocation skills will make more accurate predictions of future yields and profitability, and thus will make more efficient adoption decisions. Similarly, imperfect information concerning new technologies may result in risks associated with innovation adoption, likely raising the possibility of committing errors. Nevertheless, while the acquisition of information shifts the probability of adoption, it does not constitute a prerequisite for adoption. For instance, in the late adoption stages, as is the case of organic farming practices, farmers may simply imitate their neighbours.

In addition, farmers are more likely to gather technical information from various sources. Kihlstrom notes that the producer's decision to gather information is more complicated when information is available in increasing degrees of reliability at increasing costs. Hence, the determinants of the adoption decision may differ with the channels of information dissemination [a theory supported by Wozniak, and Gervais, Lambert, and Boutin-Dufrense]. In this context, following Feder and Slade and Jensen, we distinguish between active and passive sources of farming-related information gathering.

The former refer to the case wherein the farmer acquires farming information incidentally from various information media (e.g., newspapers, television, and radio; visits to agricultural product fairs and shows; sporadic attendance at seminars, meetings, or demonstrations) and from agricultural input suppliers. Passive sources refer to the case wherein the farmer acquires farming-related information via periodic contacts with public or private extension agents.

Regarding technology adoption, farmers often choose to adopt only parts of an innovation rather than the entire package, or they opt to apply the new technology only to one portion rather than to the whole farm. In the case of organic farming adoption, it is common practice for farmers to convert only a portion of the farm (or only one of the farm's activities) to organic. Therefore, a useful criterion regarding organic farming adoption is the intensity (or degree) of adoption with respect to the size of the farm operation. For the purpose of this analysis, we separate organic farming adoption intopariia/ (when organic techniques are applied only to a portion of the farm's total acreage) and total (when the whole farm is converted to organic).

The parameter estimates of the system of equations defined indicate only the direction of the effect of each explanatory variable on the response probabilities of the information acquisition and of total or no technological adoption. The exact effect of each explanatory variable on the individual probabilities of the three response variables requires computing the marginal effects of the regressors.

The data used in this study come from a broader survey of the structural characteristics of the agricultural sector on the Greek island of Crete, financed by the Regional Directorate of Crete in the context of the "Regional Development Program 1995-99". The sample consists of 237 randomly selected multi-crop farms located in the four major districts of Crete during the 1996-97 period. Detailed information about production patterns, input use, average yields, gross revenues, and structural characteristics of the surveyed farms were obtained via questionnaire-based field interviews. Our choice of the variables included in the information acquisition and organic farming adoption decision equations as explanatory variables is based on previous empirical evidence reported in the literature as well as on the availability of the relevant information arising from our sample survey. Based on the primary data collected, the factors affecting the farmer's information acquisition and adoption decision processes are classified into four categories: (a) farmer's personal characteristics, (b) economic variables, (c) institutional factors, and (d) environmental conditions.

From the total 237 farms in the sample, 75 (31.6%) have adopted organic farming methods into a part of their holdings, while only 44 (18.6%) have completely converted their land into organic production. The remaining 118 farms (49.8%) have not adopted any organic farming techniques.

The average years of education for farmers adopting organic farming methods is 11.4 years (10.4 and 13.0 years for partial and total land adopters, respectively), which is considerably higher than the corresponding figure for those farmers using conventional farming practices (7.7 years). Highly educated farmers may acquire technical information more easily, as their capacity to assimilate information from various sources is assumed to be greater. Educated farmers read technical bulletins and innovation-describing leaflets more than their less educated counterparts, presumably because they find it profitable to do so. As a human capital variable, education is also expected to positively affect the efficiency of adoption.

Better educated farmers are adopting profitable new technologies faster since the associated payoffs from innovations are likely to be greater and the risk is likely to be smaller. Indeed, better educated farmers are able to discriminate between promising and unpromising ideas, and hence are less likely to make allocative mistakes. Thus, one would expect a farmer's education level to be positively correlated with the decision to adopt or not adopt organic farming and the information acquisition process.

The average age of the head of the household is 42.3 and 41.3 years for partial and total organic land adopters, respectively, whereas the corresponding age for nonadopters is 55.6 years. Age is highly correlated with experience, and therefore its effect can be considered as the composite effect of farming experience and the planning horizon. Experience, in turn, provides increased knowledge about the environment in which decisions must be made.

Thus, experience may serve as a substitute for information, or at least it may modify the decision set for which information is sought. The impact of the farmer's age on technological adoption is less clear. Considerable farming experience is expected to affect adoption positively, but younger farmers with longer planning horizons may be more likely to invest in new technologies. On the other hand, if farmers are not faced with significant capital constraints and take future generations' welfare into account, the primary effect of age is likely to increase the probability of adopting technological innovations.

Average farm size is 3.65 ha for non-adopters, 4.04 ha for partial land adopters, and 3.07 ha for total land adopters. Farm size may also affect both information acquisition and technical adoption. The direction of these effects, however, is less clear. Larger farms have a greater potential to convert a portion of their land to organic farming. This is partly explained by the associated high costs involved in organic conversion (e.g., developing new markets and distribution channels, financing new activities) and risk considerations. Yet, larger farms may have less financial pressure to search for alternative ways to improve their income either by switching to a different farming technology or by seeking out technical information. Small farms generally adopt more labour-intensive technologies as they use relatively more family labour which can have a low opportunity cost. In this context, conversion to organic farming may serve as a good alternative for smaller farms, as it requires more on-farm labour than conventional farming practices.

The Herfindahl index, which measures a farm's specialisation, has an average value of 0.767 for non-adopting farms, indicating a greater degree of specialisation compared to partial and total adopters, whose respective average values are 0.494 and 0.410. Indeed, specialised farms have fewer requirements for technical assistance, and thus for information gathering, as their expertise is continually improved over time. Nevertheless, production specialisation may positively or negatively impact a farmer's decision to adopt technical innovations. Farmers growing a single crop are faced with a higher risk associated with future yields and thus farm income, which in turn induces a lower level of adoption.

Off-farm income is hypothesized to provide financial resources for information acquisition and to create incentives to adopt new technologies as the opportunity cost of time rises. On the other hand, the level of off-farm income may not be exogenous, but instead influenced by the profitability of farming itself, which in turn depends on adoption decisions.

However, in our survey, off-farm income arises mainly from non-farm business activities (e.g., tourism) and from employment in other non-farm sectors (e.g., public administration, construction work). Given the difference in skill requirements for these jobs, farm and off-farm income may be realistically assumed to be noncompetitive. Thus we can assume the level of off-farm income could be largely exogenous to adoption decisions. It appears that adopters have higher off-farm income than those who have not adopted any organic farming practices. Specifically, non-adopting farmers receive on average 640 Euros (euro)/year from off-farm sources, partial land adopters 954 euro/year, and total land adopters 980 euro/year.

The percentage of farmers contacting extension agents either from public or private agencies is 88.6% for total land adopters and 70.7% for partial land adopters. In contrast, only 16.9% of non-adopting farms receive extension services from either source. This difference is not explained by the number of extension outlets in the area and their distance from the farm, despite the differences among farm groups.

Similarly, farmers actively seeking information are mainly among the organic adopters (partially or totally), with only 11.9% of non-adopters actively seeking farming information. This finding is partly (but not completely) explained by farms' distance from urban areas. There are transaction costs associated with actively searching for relevant information. When these costs are lower, the probability of actively seeking information will be increased. The distance from urban areas is assumed to capture these transaction costs of acquiring relevant information.

These costs are expected to be larger when farms are farther from urban areas. Unfavourable environmental conditions in the area where a farm is located may also increase the risk of future yields, and thus decrease farmers' propensity to adopt organic farming practices. Adverse environmental conditions (as measured by the aridity index) seem to negatively affect farmers' propensity to adopt organic farming methods.

This finding is not surprising given that organic farming is a low input farming mode more vulnerable to adverse environmental conditions. Subsidies 19 received in the context of CAP may reduce the financial pressure on the farm, and consequently may be expected to positively affect adoption decisions.

As non-adopters are receiving the sum of only 652 euro/ year on average, whereas partial and total land adopters are enjoying 1,001 and 1,459 euro/year, respectively. Finally, we can assume that farmer awareness about environmental degradation may also induce organic farming adoption. As noted by McCann et al. and Laajimi and Albisu (2000), organic farmers express great concern about environmental problems linked to agriculture. Specifically, 79.5% of the total land adopters express a great concern about environmental degradation, a considerably higher value than the corresponding 16.1% for non-adopters.

For the estimation of the trivariate ordered probit model, we implement the GHK algorithm with 100 repetitions. Focusing first on the lower part, it can be seen that the estimated correlation coefficients lend support to the hypothesis that the decisions of farming information acquisition and organic farming adoption are correlated. Specifically, the positive interaction between the two modes of information acquisition implies the likelihood of acquiring farming information periodically from public or private extension agents is positively related to the likelihood of acquiring this information actively from other sources.

In addition, the positive correlation found between each type of information acquisition and the decision to adopt organic farming methods suggests that information-exposed farmers are more likely to be organic adopters. Statistical testing further supports these findings, as each individual correlation coefficient is statistically significant at either the 1% or 5% significance level.

Moreover, the hypothesis of no pairwise correlations between the two types of information and organic adoption is rejected at the 1% significance level. Therefore, information acquisition should not be treated as an exogenous variable when estimating a model for adoption. It should also be noted that the model correctly predicts 74.41% of individual probabilities. The maximum-likelihood (ML) coefficient estimates of equations (4a)-(4c) are reported in the upper part. The majority of the estimated parameters are statistically significant at least at the 5% level. These estimates, however, have a limited interpretation due to the discrete nature of the dependent variables; therefore, the marginal effects of the explanatory variables were computed.

The factors raising a farmer's probability of acquiring farming information (via periodic contacts with public or private extension agents) are the education level and the number of available extension outlets. Specifically, the marginal effect for education suggests, ceteris paribus (i.e., holding all other variables constant at their sample means), a farmer with one more year of education than the average level in the sample has a higher probability (by an amount of 0.147) of acquiring farming information periodically via extension. Similarly, the marginal effect of the farming extension outlets available implies, ceteris paribus, farmers with access to one more outlet than the average number of outlets in the sample have a higher probability of acquiring farming information periodically (by an amount of 0.108).

The factors reducing a farmer's likelihood of acquiring extension-based farming information are the farmer's age, the farm's specialisation, the farm size, and the farm's distance from extension outlets. Hence, a unit increase in the farmer's age, the farm size, or the farm's distance from extension outlets, or a 1% increase in the farm's Herfindahl index, appear to reduce ceteris paribus the probability of acquiring farming information, by 0.092, 0.050, 0.085, or 0.036, respectively. Finally, off-farm income does not appear to exert any significant influence on the farmer's decision to acquire extension-based information.

The same factors, however, influence differently the farmer's decision to actively seek farming information via other media sources. Education and information availability (as measured by the farm's proximity to urban centres, where presumably the chances of exposure to all kinds of information are higher) increase the probability of a farmer acquiring farming information sporadically via various media sources.

However, age, off-farm income, and farm specialisation negatively influences the same decision; moreover, farm size does not appear to have any significant impact on this decision. Our approach has the additional advantage of disentangling the role of the factors determining both the fanning information acquisition and the organic adoption decisions. Specifically, the marginal effect on the adoption decision of variables that jointly determine the farmer's decision to seek information and to adopt organic production is the combination of an indirect and a direct component, each one reflecting the variable's effect in the information acquisition and the adoption processes.

Thus, the combined effect for age is negative, implying a 0.096 decrease in the probability that a farmer who is a year older than the average age in the sample will convert partially to organic farming. Moreover, this decrease is primarily due to the lower likelihood of older farmers to acquire farming information either periodically via extension or sporadically via various media. Indeed, a unit increase in a farmer's age reduces directly the probability of partial adoption by 0.036, and it also reduces the same probability indirectly by 0.06 through the negative effect of age on the likelihood that a farmer will seek farming information. Similar interpretation holds for all marginal effects involving variables affecting both the information acquisition and the adoption decisions.

With respect to the factors shaping the farmer's decision to adopt partial organic farming, education, subsidies received, environmental awareness, and farming information gathered either passively or actively raise the probability of partial organic farming adoption. In contrast, the farm's specialisation, farmer's age, and less favourable climatic conditions (as measured by the aridity index) decrease the same probability. Farm size, off-farm income, and the farm's proximity to urban areas do not significantly affect the adoption decision of partial organic farming.

The identical determinants influence the farmer's decision for total organic conversion in the same fashion; however, the magnitude of their marginal effects differs. The marginal effects of "extension contacts" and "active information" are larger in the decision of a farmer to become fully (rather than partially) organic; this implies the farmer's decision to gather farming information (either passively or actively) raises the probability more for full rather than partial organic farming adoption. In contrast, the marginal effects of the remaining determining factors-age reduction, education, climatic conditions, subsidies, farm specialisation, and environmental awareness-suggest these factors heighten the probability of the farmer to adopt partial organic farming.

## Policy Implications

The positive correlation found between the two modes of information acquisition implies that (public and private) extension and other media are complementary sources of farming information. Moreover, younger, better educated farmers seem more likely to both acquire farming information and adopt partially or fully organic farming technology in their operations. Highly educated farmers acquire technical information more easily, as their capacity to assimilate information from various sources is assumed to be greater.

On the other hand, as a human capital variable, education does positively affect the efficiency of adoption. Similarly, age as a proxy for experience can serve as a substitute for farming information, whereas younger farmers with longer planning horizons may be more likely to invest in new technologies. The farm size appears not to influence most of the decisions examined in this study. It only affects the farmer's decision to seek extension information: owners of larger farms seem less likely to obtain farming information via extension contacts.

Off-farm income influences neither the information acquisition nor the adoption decisions. With off-farm income sources often viewed as a means of finance for information acquisition and new technology adoption, it has been argued that organic farming is an activity particularly favoured by farm operators with considerable off-farm finances. Environmental awareness appears to have a larger impact on the decision for partial rather than full organic conversion. This finding is not unexpected, given the higher risks associated with converting the entire farm (rather than one portion) to organic.

The larger marginal effect of subsidies on the decision for partial rather than full organic conversion is also interesting; in fact, it may be contrasted with EU plans to convert a sizable portion of European farming into organic production by implementing CAP measures which are almost exclusively subsidy-driven. Moreover, favourable climatic conditions seem to be a consideration in the farmer's decision for organic technology adoption, which supports similar results reported by McCann et al. (1997).

The policy recommendations implied by these results can be summarised in terms of the diffusion strategies planning authorities may wish to pursue with respect to organic farming adoption.

First, active and passive information sources have different audiences and certainly different costs for the decision makers. Policies and practices of information providers should reflect the specific characteristics of potential adopters to enhance the return of information dissemination activities and better serve farmers' needs. Active information providers should target farmers with low levels of off-farm income who are less specialised in their farming activities and whose operations are close to urban centres. Extension services can target small, highly diversified farms with higher educated operators who have a greater capacity for processing and decoding technical information.

Moreover, if policy makers wish to encourage partial (or gradual) organic adoption, then policy measures should address:

(a) the improvement of farmer education,

(b) the retirement of aging farm operators,

(c) the development of farming information channels and networks (including extension services),

(d) the cultivation of environmental considerations among farmers,

(e) the encouragement of multi-output oriented farms, and

(F) the advancement of public and private extension services.

This study suggests that a farmer's decision to adopt new technologies (specifically organic farming) should not be studied separately from the decision to acquire farming information. To explore this issue, we specified a structural probit model in which the farmer's decision to acquire farming information via different sources and the decision to adopt a new technology are correlated. Our model is applied to a cross-sectional sample of Cretan farmers. The empirical results show that acquisition of farming information and organic adoption are indeed correlated decisions.

Moreover, the sources via which farmers gather farming information appear to be positively related. This finding implies that different sources play a complementary role in the farmer's decision to gather information. Policy insights derived in the context of this study suggest that measures to promote the adoption of organic farming techniques should be primarily structural rather than subsidy-driven.

Specifically, our findings indicate that organic farming adoption would be mainly influenced by policy measures which encourage the retirement of older farmers; improve farmers' education, environmental awareness, and information channels and networks; encourage farm output diversification; advance extension services; and organise workshops and round table meetings among farmers and rural stakeholders. Further, the development of extension services (public or private) appears to be pivotal if a strategy of organic adoption is to be pursued.

Within the context of EU Rural Development Regulation 1257/1999 as amended by Regulation 1783/2003, several measures (2nd Pillar) can be undertaken toward the restructuring of farm operations to diminish their difficulties in converting to organic. In addition, the recent CAP Reform toward decoupled farm incomes can also provide a positive framework for the future development of organic agriculture throughout the EU.

However, since Member States have different options regarding implementation, the degree of decoupling and the use of national envelopes will have an impact on organic farming. Hence, Member States wishing to support organic agriculture should consider the factors affecting farmers' decisions when implementing the new rules. The existence of national envelopes gives the opportunity to each Member State to deal with its own peculiarities and structural difficulties regarding farm operations so that organic agriculture can be expanded.

## Organic Agriculture for Market- and Risk-Based Regulation

In hindsight, the year 1990 was a crossroads in environmental and regulatory policy generally and in the regulation of food specifically. And two different narratives capture the competing regulatory choices presented at that crossroads. The first, a conventional narrative, finds mounting evidence of what many regard as the dominant avenues of regulatory reform in environmental policymaking: market-based incentives and risk-based decisionmaking. The second, alternative narrative finds the signs of a much more fundamental reform project.

Under the conventional narrative, 1990 is often remembered for amendments to the Clean Air Act that inaugurated the country's largest experiment with constructed environmental markets—a nationwide "cap-and-trade" sulfur dioxide market now feted not only as a template for a worldwide market in greenhouse gases but also as a model for the development of more far-reaching markets in ecosystem services. The year is also recalled for the qualified endorsement given quantitative and comparative risk assessment by the Environmental Protection Agency's Science Advisory Board (SAB). Thereafter, the SAB's report would be cited frequently for evidence that risk-based environmental regulation was an idea whose time had come.

In the regulation of food, it was easy to see in 1990 the influence of these conventional reforms, especially the prevalence of risk-based regulation. Two highly visible controversies over food—one domestic and one international—were largely framed and resolved in terms of risk assessment. Domestically, 1990 marked the conclusion of a regulatory battle over the pesticide daminozide (known by its trade name, Alar) in which the Environmental Protection Agency (EPA), the Uniroyal Corporation (Alar's chief manufacturer), and environmental groups each contested the safety of Alar through opposing risk-based conclusions involving competing exposure models and dose-response relationships.

Internationally, the year also saw battle lines drawn over hormone-treated beef in a dispute between the United States, where the addition of hormone growth implants in livestock was (and still is) common, and the European Community (EC), which took final steps to implement its ban on imported, hormone-treated meat adopted the previous year.

The battle over beef hormones pitted a risk-based approach to environmental issues in international trade (the United States' position) against a precautionary approach (the EC's position). And, within the dispute resolution framework of the World Trade Organisation (WTO), the United States' position would prevail, not only in the WTO's eventual decision in 1997 on bovine growth implants, but also in a 2006 dispute between the United States and Europe over the importation of genetically modified agricultural products.

But in the regulation of food, the alternative narrative is needed to capture fully other events that also transpired in 1990. Broadly speaking, this narrative emphasizes the emergence of a cause-based approach to environmental reform that seeks fundamental changes in production systems or human behaviour to prevent such environmental harms from arising in the first place. In a weak sense, Congress captured this alternative approach in the Pollution Prevention Act of 1990, under which EPA was to promote pollution prevention through various information-based measures and by encouraging source-reduction efforts by industry. But this alternative approach was captured more strongly that year by developments in "alternative agriculture"—which for purposes of this Chapter includes "organic agriculture."

The year 1990 was especially important in the development of federal policy on organic agriculture. In April of that year, the National Academy of Sciences co-sponsored a workshop on research initiatives in "sustainable agriculture," following up on a widely noted National Academy publication from one year earlier, Alternative Agriculture, which found evidence that non-conventional agricultural systems could markedly reduce environmental degradation without significant loss in agricultural productivity.

Perhaps influenced by these events, Congress passed the Organic Foods Production Act of 1990 (OFPA), as part of the 1990 Farm Bill. The OFPA was, it is safe to say, ironic. It delegated the task of promulgating a uniform set of standards for organic farmers, the National Organic Program (NOP), to the United States Department of Agriculture (USDA), the historical and administrative epicentre of conventional agriculture in Washington, D.C. Once established, the NOP would allow organic farmers to market their products under the nationwide seal, "USDA Organic."

After stumbling badly in its initial attempt to implement the delegation, the USDA finally promulgated regulations for the NOP about a decade later.

The incorporation of organic agriculture within federal environmental and agricultural policy was significant. In facilitating this approach to agriculture, the OFPA brought full circle the deeper reform project suggested by Rachel Carson almost four decades earlier upon the publication of her widely influential book, Silent Spring. Several accounts of Silent Spring link its publication to legislative reforms of the nation's central pesticide statute, the Federal Insecticide, Fungicide, and Rodenticide Act (FIFRA). But Carson argued for something more fundamental than a broader accounting of the health and environmental effects of pesticides. Carson concluded Silent Spring with a chapter entitled "The Other Road," in which she called on farmers and agricultural researchers to eschew what she termed the "arrogance" of the "control of nature," and instead to develop solutions to agricultural pests "based on understanding of the living organisms they seek to control, and of the whole fabric of life to which these organisms belong."

Unlike regulatory reform based merely on quantifying and prioritizing risks, the other road marked by the OFPA led toward a fundamentally different form of agriculture. And unlike the artificially constructed markets in pollution control established in the 1990 Clean Air Act amendments, the OFPA sought to link this different form of agriculture directly into the nation's broader consumer markets.

And it succeeded. For all the attention given the artificial market for sulfur dioxide allowances under the Clean Air Act, its annual market value in 2004 was approximately $4 billion). In contrast, sales of organic food in 2004 were approximately $15.4 billion, up from $12.9 billion in 2003. Organic foods have for years constituted the fastest growing segment of the agricultural market. The OFPA marked a road that has been taken.

After recounting briefly the basic contours of organic agriculture and USDA's development of the NOP, this Chapter squarely addresses the continuing challenge to the NOP's legitimacy. Specifically, this Chapter responds to attacks—typically framed in the language of risk assessment—that organic agriculture cannot be squared with sound public policy or with what its attackers frequently describe as sound science.

The central thesis of this Chapter is that detractors of organic agriculture too often rest their claims to sound science with arbitrarily narrow framing devices that mask the full range of benefits that organic agriculture may offer. In that regard, the battles over organic agriculture are not unique.

In January 2006, the White House Office of Management and Budget (OMB) sought to hardwire just such a limiting rubric across the entirety of the regulatory landscape by proposing, in the name of sound science, a national "risk assessment bulletin."

In December 2006, a panel of the National Academy of Sciences, after conducting an eleven-month peer review, took the unprecedented step of returning the proposal to OMB as "unscientific" and "unjustified." This Chapter discusses how the artificially limited perspectives of the rejected OMB proposal reflect just the sort of arbitrary argumentation too often marshaled in defence of conventional agriculture and against organic foods.

## Organic Agriculture and the National Organic Program

Organic agriculture can be understood on three levels, each one of which can be helpful to policy and legal analyses. At its broadest level, organic agriculture reflects a set of ethical positions—toward the environment, toward socioeconomic justice, and toward animal welfare—as well as a set of agricultural methods. The International Federation of Organic Agriculture Movements speaks of four overarching principles of organic agriculture—of health, ecology, fairness, and care—and stresses their applicability "to agriculture in the broadest sense, including the way people tend soils, water, plants and animals" and to the way people "relate to one another and shape the legacy of future generations."

It is not hard to catch in such meta-principles the philosophical, as well as technical, aspects of the "other road" that Rachel Carson propounded. And, while still describing organic agriculture at this broadest level, it bears mention that organic agriculture also has social and organisational dimensionality; there are hundreds of regional and local organic industry trade groups, each often having slightly varying standards. Moreover, at the international level, in addition to the standards set by the private International Federation of Organic Agriculture Movements, there are also standards for organic food products set by the Codex Alimentarius Commission, an entity operating under the United Nations Food and Agriculture Organisation and the World Health Organisation. Thus, broadly speaking, organic agriculture has all the earmarks of a self-generating, self-policing private organisation centred on core norms that might simultaneously deliver both private goods and public beneficial externalities.

At a second, narrower level, organic agriculture can be understood as one of several "alternative" agricultural systems. All of these systems share the goal of reducing, if not eliminating, reliance on such chemical or artificial agricultural inputs as conventional pesticides and fertilizers. There are roughly four alternative methods most often suggested as substitutes. First, there are cultural methods of disease, weed, and pest control, such as crop rotations and staggered planting dates. Second, there are biological control methods, such as the release of "beneficial" predatory or parasitic insects that can attack pests. Third, there are "biorational" pest control measures such as pheromone-baited traps. And fourth—although not an option embraced by organic agriculturalists—there is "integrated pest management," (IPM) a "decisionmaking system designed to use all 'suitable' pest control techniques, including chemical pesticides, to keep pest populations below economically injurious levels while satisfying environmental and production objectives."

The National Academy of Sciences spoke of alternative agriculture generally when it concluded in 1989 that "[w]ider adoption of proven alternative systems would result in even greater economic benefits to farmers and environmental gains for the nation." Finally, there is the legal regime governing organic agriculture developed under the Organic Foods Production Act. In the late 1980s, against the backdrop of legislation in over twenty states regulating organic food labelling in different ways, the organic food industry petitioned Congress for legislation that would create a set of national standards for organic products.

The OFPA, from the point of view of regulatory design and administrative law, was strikingly innovative. At the same time that alternatives to traditional command-and-control regulation such as risk-based decisionmaking and market-like incentives were drawing so much attention, the OFPA created a system that could tie public environmental and ethical values into existing, real markets; that informed the development of governmental organic standards with input from a National Organic Standards Board composed of nongovernmental representatives from different facets of the organic industry; and that centred regulatory compliance on a system of approved private-sector certification rather than a large federal bureaucracy.

The OFPA is a marketing-oriented statute designed to regularize what was at the time a potentially confusing Babel of competing standards with an official federal "organic" label. Not only was a federal label thought useful in promoting consumer confidence in the growing organic industry within the United States, but it was also viewed as helpful in facilitating trade in "a potentially lucrative international organic market."

For this reason, the OFPA's primary administrative delegation to develop the National Organic Program was to the USDA's Agricultural Marketing Service (AMS). But because the AMS had little direct experience with organic agriculture, its decisions were to be informed by recommendations from organic industry representatives serving on the National Organic Standards Board (NOSB). Although USDA was slow even to propose establishment of the NOP, by 2002 final rules had been implemented that allowed, among other things, qualifying products to use the official USDA "organic" label in conjunction with terms such as "100 percent organic" or "organic."

By two measures, the NOP has been successful. First, in its definition of organic farming, even a business-oriented institution like the AMS recognised that, broadly speaking, organic agriculture was more than simply the application of certain techniques. True to the movement's overarching principles, organic farming was defined under the OFPA as "[a] production system that is managed in accordance with [OFPA regulations] to respond to site-specific conditions by integrating cultural, biological, and mechanical practices that foster cycling of resources, promote ecological balance, and conserve biodiversity."

But second, and as importantly, the national system of organic certification has contributed to the underlying growth of the organic food market, both domestically and internationally. The Congressional Research Service reported in 2006 (based on 2004 data) that over 11,000 certified organic crop, livestock, and handling operations were involved with over 2 million acres of certified organic cropland and pasture/rangeland in a domestic market that maintained an astonishing annual rate of growth of 20 percent beginning in 1990.

By one account, the domestic sale of organic foods might exceed $23 billion by 2010, or 3.5 percent of total U.S. retail sales (up from $10.4 billion in sales in 2003, accounting then for only 2 percent of total U.S. sales). Internationally, as of 2005 organic food production is reported to occur in 110 countries, on over 558,000 farms cultivating over 64 million acres, with global sales maintaining an annual expansion rate of 7-9 percent and having already exceeded a total of $25 billion.

Following the United States' lead in 1990, both Europe and Japan similarly adopted national standards for organic products. To the extent the OFPA was premised on the contribution national standards might make to market growth, the NOP has been successful. But there have also been signs that USDA implementation of the NOP has misunderstood the full sweep of organic agriculture's principles. The clearest sign occurred in 1997 when the NOP was first proposed. Then, the USDA had included the use of genetically modified organisms among its range of acceptable "organic" practices over the strong objection of the NOSB. The USDA's reasons for doing so were defensive rather than descriptive. The agency pointed to no evidence that organic farmers actually used genetically modified organisms, or that such use would be consistent with the broader principles of organic agriculture, or even that an independent scientific inquiry had led the agency to believe that genetically modified products were either beneficial or free of dangers.

Rather the agency principally explained that, because the United States was otherwise the leading advocate for genetically modified agricultural products, it was concerned that excluding the use of genetic engineering from the NOP would cause its foreign trading partners to believe that the federal government was concerned with the safety of food grown with some use of genetic engineering. The public response to the proposed rule is legendary. By one account, public opposition was so enormous—"it was one of the largest [public responses] in the history of the federal government" —that USDA was twice required to extend the comment period to accommodate the sheer number of critical written submissions. By another, more than 275,000 comments, almost all opposing USDA's proposed rule, were received. In the end, USDA retracted the proposal and the final NOP prohibited the use of genetic engineering as an approved practice.

In doing so, the agency stated that the OFPA was primarily a marketing statute and that as the public overwhelmingly views "organic' foods not to have been grown with genetic modification techniques, excluding them was necessary to preserve the meaningfulness of the USDA's "organic" label. As a matter of statutory interpretation, the USDA's final position was unquestionably consistent with the OFPA. But within

USDA's explanation lie shades of a less defensible conclusion: that organic agriculture, including the non-utilisation of genetically modified processes, gains its legitimacy only within the realm of public perception and could not survive the scrutiny of full-fledged scientific inquiry.

The American Crop Protection Association, an organisation that supports the use of genetic modification in farming, came close to stating this explicitly:

> *"The [ACPA] can accept the exclusion of modern biotechnology from organic production as an 'excluded method' only with the clear understanding that the organic designation is in no way an indication of safety or quality but is rather a marketing standard."*
>
> *In the remainder of this Chapter, I address this inference.*

# Bibliography


Adams, W. M. and Jeanrenaud, S. J.:*Transition to Sustainability: Towards a Humane and Diverse World.* Gland, Switzerland, 2008.

Barnett, J. : *The Fractured Metropolis, Improving the New City*, New York, Harper Collins, 1995.

Blewitt, J.: *Understanding Sustainable Development*, Earthscan, London, 2008.

Carley, M., P. Jenkins : *Urban Development and Civil Society, the Role of Communities in Sustainable Cities*, Sterling, VA, Earthscan Publications, 2001.

Cronon, W. : *Nature's Metropolis, Chicago and the Great West*, New York, W. W. Norton, 1991.

Daly, H. & J. Cobb, *For the Common Good: Redirecting the Economy Toward Community, the Environment and a Sustainable Future,* Beacon Press, Boston, 1989.

Fowler, E. P. and D. Siegel : *Urban Policy Issues, Canadian Perspectives, Toronto,* Oxford University Press, 2002.

Hawken, P, Lovins, A.B. & L.H.: *Natural Capitalism: Creating the next Industrial Revolution.* Rocky Mountain Institute, Snowmass, USA, 1999.

Huesemann, M.H., and J.A. Huesemann, *Techno-Fix: Why Technology Won't Save Us or the Environment,* New Society Publishers, Gabriola Island, Canada, 2011

Katz, P., V. J. Scully : *The New Urbanism, Toward an Architecture of Community,* New York, McGraw-Hill, 1994.

Kaza, Stephanie, and Kenneth Kraft : *Dharma Rain, Sources of Buddhist Environmentalism,* Shambhala, Boston, 2000.

Kellert, Stephen R. and Timothy J. Farnham : *The Good in Nature and Humanity, Connecting Science, Religion, and Spirituality with the Natural World,* Island Press, Covelo, California, 2002.

Krebs, C.J.: *Ecology: the Experimental Analysis of Distribution and Abundance,* Benjamin Cummings, Sydney, 2001

Marris, P. : *Community Planning and Conceptions of Change,* London, Boston, Routledge & Kegan Paul, 1982.

Perrings, C. : *Sustainable Development and Poverty Alleviation in Sub-Saharan Africa,* Botswana, New York, St. Martin's Press, 1996.

Sassen, S. : *Cities in a World Economy, Thousand Oaks,* Calif., Pine Forge Press, 2000.

Selin, Helaine : *Nature Across Cultures, Views of Nature and the Environment in Non-Western Cultures,* Kluwer Academic Publishers, Lancaster, U.K. 2003.

Simonds, J. O. : *Garden Cities 21, Creating a Livable Urban Environment,* New York, McGraw-Hill, 1994.

Smith, M. P. : *Pacific Rim Cities in the World Economy,* New Brunswick, USA, Transaction, 1989.

Taylor, Bron, and Jeffrey Kaplan : *The Encyclopedia of Religion and Nature,* Thoemmes Continuum, London & New York, 2005.

Tirosh-Samuelson, Hava : *Judaism and Ecology, Created World and Revealed Word,* Harvard University Press, Cambridge, 2002.

Tucker, Mary Evelyn : *Worldly Wonder, Religions Enter Their Ecological Phase,* Open Court Publishing Company, La Salle, Illinois, 2003.

Tucker, Mary Evelyn, and John Grim: *Worldviews and Ecology: Religion, Philosophy, and the Environment.* Orbis Books, Maryknoll, N.Y. 1994.

Uhlir, P. F. : *Scientific Data for Decision Making Toward Sustainable Development,* Washington, D.C., National Academies Press, 2003.

Van der Ryn, S. and P. Calthorpe : *Sustainable Communities, a New Design Synthesis for Cities, Suburbs, and Towns,* San Francisco, Sierra Club Books, 1986.

Veitch, James : *Can Humanity Survive, The World's Religions and the Environment,* Awareness Book Company Ltd, Auckland, New Zealand. 1996.

W.A., Western : *Liveable Neighbourhoods, a Western Australian Government Sustainable Cities Initiative,* Perth, Australian Planning Commission, 2000.

Walter, B., L. Arkin : *Sustainable Cities, Concepts and Strategies for Eco-city Development,* Los Angeles, CA, EHM Eco-Home Media, 1992.

# Index


## A

Acquisition, 136, 137, 138, 139, 140, 141, 142, 143, 144, 145, 146, 147.
Adoption, 38, 43, 44, 59, 135, 136, 137, 138, 139, 140, 141, 142, 143, 144, 145, 146, 147, 153.
Air Pollution, 41, 77.

## B

Biodiversity, 8, 25, 26, 28, 30, 34, 37, 41, 47, 62, 73, 104, 106, 114, 115, 154.
Biogeographic Evaluation, 107.
Butterflies, 25, 27, 37, 75, 77.

## C

Carbon, 6, 12, 18, 64, 66, 67, 68, 116, 118, 119, 140.
Certification Issues, 23.
Chemical Contamination, 109.
Consequences, 17, 32, 74.
Consumer Safety, 120, 122.
Cover Crops, 15, 40, 67, 69, 70, 71, 72, 132.
Crop Diversity, 4, 12, 113.
Crop Rotations, 2, 66, 72, 152.

## D

Deep Ecology, 45, 46, 47, 48, 49, 50, 51, 52, 53, 55.
Drainage, 2, 64, 69, 71.

## E

Earthworms, 25, 28, 66.
EEffectiveness, 16, 31, 136.
Energy Efficiency, 6, 114, 120.
Environmental Impact, 42, 119, 113, 117.
Equipment, 13, 67, 69, 71.

## F

Farm Size, 12, 117, 140, 143, 144, 145.
Fertility,13, 15, 16, 39, 40, 85, 116, 121, 132.
Food Safety, 19, 109.

## G

Greenhouse Gases, 67, 148.
Gullies, 69, 71.

## H

Habitats, 26, 34, 36, 37, 95, 98, 102.
Honey Bees, 74, 75, 79, 104.

## I

Identifying Organic Food, 117.
Insects, 9, 12, 13, 14, 16, 17, 18, 26, 34, 35, 36, 64, 68, 72, 73, 75, 78, 80, 104, 105, 118, 131, 153.
Invaded Ecosystems, 100.
Invasive Species, 95, 96, 97, 98, 99, 100, 103, 104, 107.

## L

Light Pollution, 77.
Livestock, 2, 3, 6, 7, 14, 15, 16, 41, 42, 44, 56, 59, 74, 87, 94, 116, 117, 120, 149, 154.

## M

Mammals, 25, 28, 34, 105, 109.
Multiple Cropping, 61, 62.

## N

Native Plants, 37, 74, 75, 98.
Nectar Corridors, 73, 76.
Nutritional Value, 62, 112, 122.

## O

Organic Agriculture, 8, 10, 110, 125, 132, 135, 148, 150, 151, 152, 153, 154.
Organic Bean Markets, 129.
Organic Beans, 128, 130, 131, 132.
Organic Certification, 18, 22, 111, 128, 154.
Organic Farming, 7, 8, 9, 10, 11, 12, 13, 14, 15, 16, 25, 26, 27, 29, 30, 32, 135, 136, 137, 138, 139, 140, 141, 142, 143, 144, 145, 146, 147, 148, 154.
Organic Farming Practices, 14, 25, 32, 117, 136, 137, 141, 142.
Organic Farms, 6, 13, 25, 26, 27, 28, 29, 30, 113, 114, 115, 118, 119.
Organic Food, 11, 18, 19, 23, 33, 38, 108, 112, 113, 114, 116, 117, 119, 120, 121, 122, 123, 124, 125, 151, 152, 153, 154.

## P

Pathogens, 70, 105, 131.
Perennials, 89, 91, 102.
Pest Management, 14, 15, 114.
Pollination Management, 72, 73, 79, 80.
Pollinator Decline, 73, 74, 75, 78.
Pollution, 3, 10, 40, 66, 77, 106, 149, 151.

## S

Sheet Mulching, 65, 81, 82.
Soil Microbes, 25, 26, 66, 67.
Succession Planting, 86, 87.

## T

Thermal Methods, 93.
Transportation, 3, 7, 114.
Tropical Climates, 56.

## U

Underutilised Crop, 62.
Urban Planning, 6.

## V

Vectors, 100, 101, 105.
Vegan organic gardening, 38.
Vegan Permaculture, 39.
Veganic Gardening, 39.
Vegetarianism, 39, 40, 42, 44.
Vegetation, 25, 29, 35, 36, 39, 58, 102.

## W

Water Issues, 71.
Weed Control, 16, 39, 89, 90, 92, 93.
Weed Pests, 130.
Wildfire Suppression, 102.
Wildlife Garden, 34.

❑❑❑